public
private
ephemeral

CERAMICS
IN ARCHITECTURE

4

A conversation with Eduardo de Miguel, Víctor Echarri and Vicente Sarrablo

Eduardo de Miguel
(Professor in Valencia)
Holder of a doctorate in architecture from the Polytechnic University of Madrid School of Architecture, Eduardo de Miguel is currently Director of Architectural Projects and Head of the Ceramic Studies Department at the Valencia School of Architecture. He has also held a fellowship at the Spanish Academy in Rome and was Visiting Fellow at the Princeton University School of Architecture. His work has received a number of distinctions, noteworthy among which are the 9th Venice Architecture Biennale and three spanish architecture Bienales

Víctor Echarri Iribarren
(Professor in Alicante)
Holder of a doctorate in architecture from the University of Navarre, Víctor Echarri is Assistant Director of the Department of Architecture and Professor in the Architectural Construction Department at the University of Alicante Polytechnic School. His research is centred on military architecture and technological innovation with ceramics and he has been responsible for publications and research projects in these areas. He combines this task with the architectural pratice

Vicente Sarrablo
(Professor in Barcelona)
Holder of a doctorate in architecture from School of Architecture of the Polytechnic University of Catalonia, Vicente Sarrablo is Professor and Head of both the Construction and Ceramic Studies Departments at the International University of Catalonia School of Architecture.
An expert in technological innovation with ceramic materials, fast-assembly industrial construction and advanced design of inflatable structures, he has worked on numerous projects, publications and in research in these fields and has received numerous awards

BETWEEN TRADITION AND INNOVATION

As a result of recent technological advances, the ceramic industry finds itself straddled between traditional techniques and applications and continuous development that is leading to surprising innovations.
What uses do you think ceramics can have beyond the areas where it has traditionally featured?

EM As a consequence of creativity and technology, we are likely to find ourselves looking at a new renaissance in ceramic materials. In my opinion, the problem does not lie in the product, but rather in our prejudice. The recent appearance of unique works where ceramic has become the leading element has helped us to realise that the fields where these materials can be applied may be unlimited; they can be worthy of use anywhere.

VS The very fact that the answer to this question is so open-ended is what makes ceramic such an interesting material for architecture. We are at a time that is ripe with possibilities. As far as interiors go, it has everything in its favour, and a way is being found for it to become a strong competitor with other materials for outdoor applications.

VE There are solutions like floating paving and curtain walls that have revolutionised the use of ceramics. Applications in urban paving are proving to have excellent qualities, although more research is still needed if porcelain stoneware is to become competitive with hydraulic and stone slab paving. We have developed a patent application for ceramic manhole covers to be used in urban infrastructure projects. Our experience in these two yeas that Ceramic Studies Departments have existed is that there are potential areas of research into solutions for façades, dropped ceilings, cable trays for electrical installations, packaged terminal air conditioning units, sustainable solutions for thermal inertia, radiant floor and ceiling cooling and heating, and so on. There are boundless possibilities for its use.

RESEARCH, INDUSTRY AND MARKET

The Ceramic Studies Departments may be considered as an important structural element, not only for its value in education but also as a source of reflection and research for the market. From the educational research perspective, what action do you think is needed for projects that are created to turn into products or techniques for application with a place in the market?

EM Reflection and research with these materials leads us to two different situations.
One where research is into new ways of applying existing materials, and the other, which is where new products are required to solve new or unique problems.
Both ways are possible and desirable; they have had brilliant results such as the New Art Gallery in Walsall by Caruso St John, in the case of work done with catalogue pieces, or the Spain Pavilion for the Expo in Aichi by FOA. This is the result of direct collaboration between manufacturers and the architects to create a new product, which will bring great prestige to the material given its uniqueness. However, this is largely an ephemeral effort in most cases because its exclusivity makes it difficult to produce commercially.
It is understandable that most manufacturers should prefer the first option. It is pleasing for them to see how their products have a wider scope of use than initially foreseen, but this can lead to a kind of complacence and lack of motivation to innovate.
Having said this, I should also say that they take risks, but in most cases only if there is a guaranteed order.

VE It is important to work in continuous collaboration with ASCER [the Spanish Ceramic Tile Manufacturers' Association] and the Ceramic Technology Institute (ITC) because of their experience in the field of research. Likewise, the role of the OTRI [Office of Technology Transfer] in universities is essential to help and promote the setting up of government-funded agreements, PETRIs [Technology Transfer Projects] and Research Projects. The fruit of this way of understanding the private enterprise-university relationship is the patent our department has been developing. We are also collaborating as partners in the ceramic sector's 4 Senses Research Project. This collaboration is vital in order for ideas generated in universities to become viable and subsequently developed by the business sector.

VS I believe there are two goals that should be met. The first is to achieve high standards in proposals by adjusting their geometric definitions, understanding the industrial processes involved and projecting their limitations for commercial acceptance. The second is that the projects should create open-ended systems to allow multiplication of solutions for their advocates who will later experiment with them, allowing them to borrow and tailor them to needs.

SUSTAINABILITY

Environmental sustainability has now become a basic issue to consider in all sectors; logically, this includes construction.
What is you opinion of the part ceramics has to play in this debate?

EM For us to speak about sustainability, we would first have to agree on the term. It is true to say that it is one of the industries making the most efficient use of the energy resources needed for production, and that is already saying a lot. But if we consider recycling of the materials produced, there may still be a great deal to be done. Although we are all aware that, because we are dealing with inert materials, it should not be difficult to make use of them.

VE Ceramic has been used as a material for thousands of years. Its sustainable nature is not under discussion, but I do understand that reflection is needed to justify its use in agreement with our natural resources. There have been a number of studies made in our country and in Great Britain in which the sustainability of ceramics is considered satisfactory compared to other materials. But we have to keep working on it. We have to support research on ceramic reuse in sustainable environmental renewal and for energy saving on façades and roofs by means of the clay jug heat loss effect, and for other techniques featuring ceramic materials.

VS I think if you take the fact that it is a material that resists the passing of time and add serious research on recycling to that, ceramic may achieve very sustainable life cycles.

FUTURE CHALLENGES

Working with students with their youth and 'clean', unobscured way of seeing things sometimes presents fresh and imaginative solutions that occasionally lead to real innovations.
What is the balance of these two years in this aspect? And, in particular, what are the challenges that your departments face in the immediate future?

EM The balance could not be more positive. There are solid proposals that often do not reach their final stage, production and marketing, because they are perceived as a risk. It would be desirable if companies with means had the courage to seriously study the viability of some of the outstanding ideas that are coming out of these departments. The greatest challenge, therefore, is to gain the confidence of the manufacturers so that the work carried out in the departments might be made into something tangible, real.
Without being self-complacent, I think the road we have taken is very good. I also think more time is needed before we can take rigorous stock of the results we have obtained.

VE I feel that setting up the Ceramic Studies Departments has awakened great interest in architecture schools. The balance could not be more positive as an educational tool. Students must put much of the knowledge acquired in the different areas such as projects, construction and systems into integrated architectural solutions. This always goes beyond the teacher's expectations owing to their capacity for innovation, more intense dedication and motivation. I think the challenges to be considered by our departments in the future are linked to research in association with companies.
A mutual enrichment is possible and this will become a driving force behind the development of architectural solutions to contemporary and global needs.

VS Some prestigious architects are showing great courage by incorporating daring ceramic pieces in their buildings. Likewise, it would be desirable if the ceramic industry corresponded boldly when it comes to considering the solutions presented by architecture students. Right now, the sector is looking but will not touch. And that is where our greatest challenge lies.

Eduardo Souto de Moura

Metro stations: Aliados, Combatentes, Casa da Música

Porto, Portugal

These stations are part of the Normetro project in the city of Porto, designed by Souto de Moura and wich began in 1997. Taking advantage of the situation he observed at the introduction of an infrastructure network in the city's fabric as an opportunity with great potential for creating relations between consolidated, disperse or planned spaces, Soto de Moura transformed the Porto metro into a landscaping operation. Therefore, regardless of the characteristics of the line – above or below ground – or of the actual stations – either in a tunnel or on the surface, the whole layout of the public space was redefined. It triggered a tentacular urban development operation with the capacity to create new opportunities in those places falling within the area that benefited from the new network. By inverting standard architectural elements, Souto de Moura created a generic list of conditions that were adaptable to the different circumstances, by means of a neutral design able to respond to different requirements. The character of the different settings of each station thus maintained a uniform line. The same conventional materials were used, while the homogeneous surface shades, marked by the use of ceramic tiles on walls and columns, give the feeling of expanse and turn each station into an outstanding civic space.

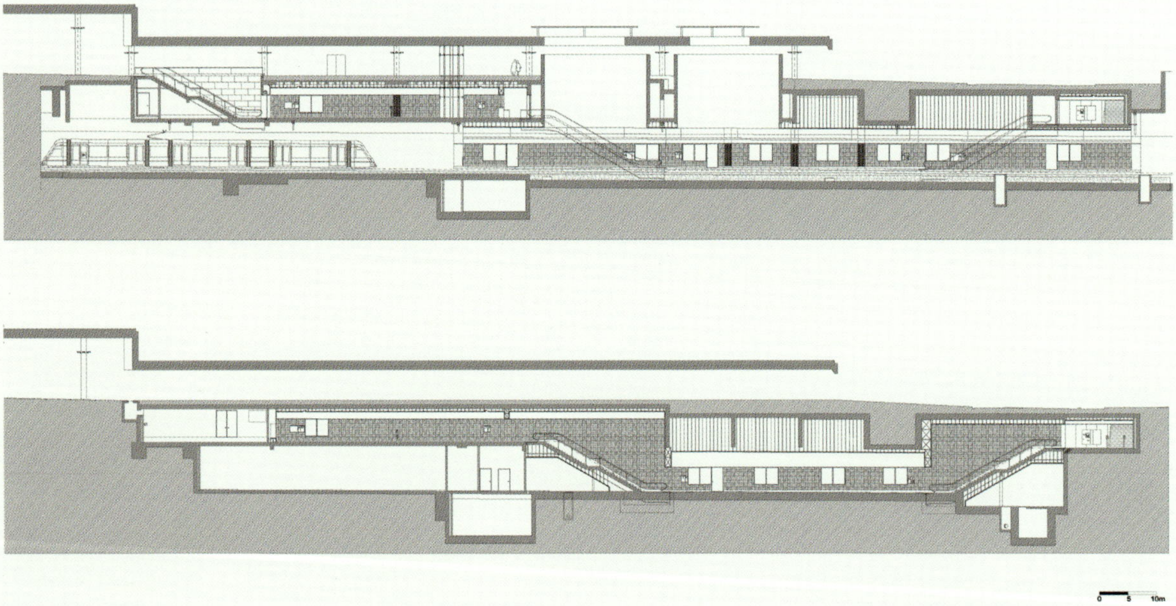

CASA DA MÚSICA STATION. SECTIONS

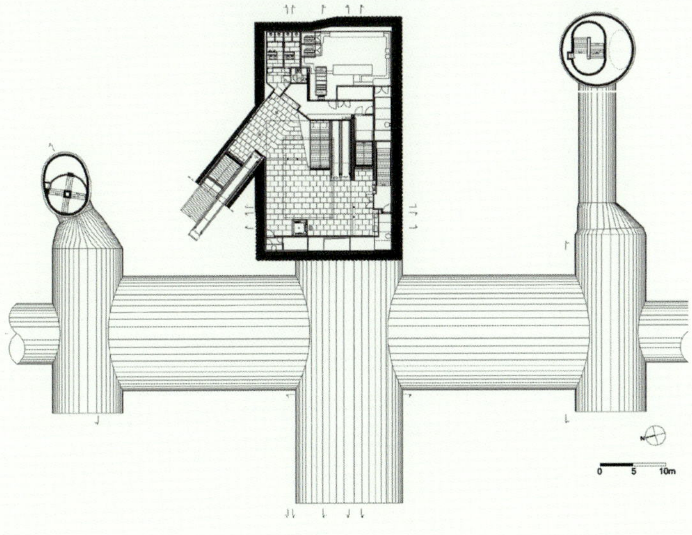

COMBATIENTES STATION ENTRANCE AND VESTIBULE

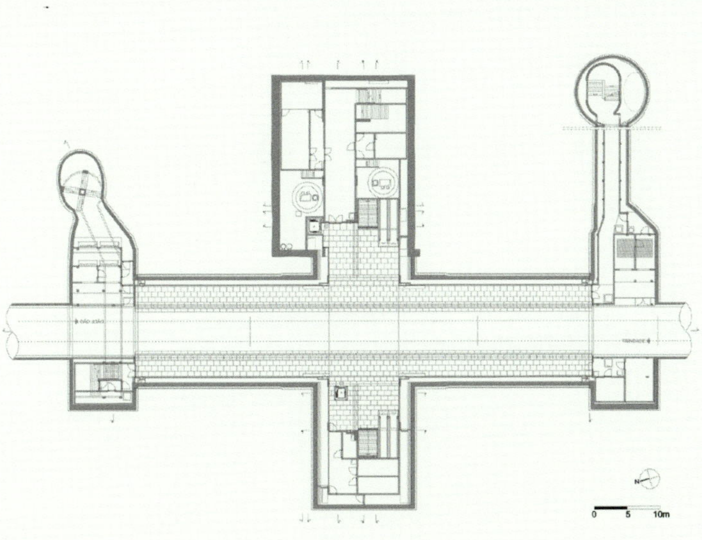

PLATFORM LEVEL

ARCHITECT: Eduardo Souto de Moura / **LOCATION:** Porto, Portugal / **PROJECT:** 1997 / **CONSTRUCTION:** 2005 / **COLLABORATORS:** Adriano Pimenta, André Campos, Ricardo Tedim, Eduardo Carrilho, Joana Pinho, José Carlos Mariano, Bernardo Durão, Diogo Crespo, Manuel Pais Vieira, Nuno Flores, Nuno Lopes, Tiago Coelho, Tiago Figueiredo, Eduardo Pereira, Pedro Chimeno, Soares da Costa / **CLIENT:** Metro do Porto / **STRUCTURE:** CENOR – Projectos de Engenharia, Lda. CJC Engenharia e Projectos, lda., COBA Consultores para Obras, Barragens e Planeamento, s.a., VIA PONTE Projectos e Consultoria de Engenharia, lda. / **HYDRAULICS:** FASE Estudos e Projectos, s.a. / **BUILDER:** Normetro ACE, Transmetro / **PHOTOGRAPHER:** Ilfa, Adriano Pimenta

CERAMIC: Glazed tile 15 x 15 cm on interior walls and columns

14

Equip XCL
Xavier Claramunt and Candi Casadevall

Casa Muntaner

Santa Margarida de Montbui, Spain

Ceramics have lately been used to spectacular effect, with special emphasis on colour and the dimensions of the covered surfaces. International exhibition centre pavilions or market roofs. Shopping centres. Emblematic buildings.
For these kinds of projects, budgets fade into the background as the more intangible goals of prestige and publicity come to the fore.
Casa Muntaner has as much in common with the previously-mentioned projects as it is completely different from them. This is a medium-sized detached family dwelling but it is a way of supporting the cause for using tiles outside of bathrooms and kitchens and surpassing the concept of tiling in a spectacular way, while keeping to the domestic setting.
Casa Muntaner is formed by using two curved retaining walls as the dwelling and then bending them over to create a roof. The exterior tiled finish is created with charcoal grey stoneware tiles over a tongue and groove ceramic backing attached to a galvanised steel understructure slightly apart from the wall in order to create a ventilated space. The external walls define the two main spans – one to the north and the other to the south – and at the point where the two parts of the roof structure, formed by bending the tiled walls over, should meet, there is a gap where the main room of the house is located.
The space created between the two outer walls is closed on both sides with walls covered with different materials to emphasis the presence and functions of the two retaining walls. Close to the end, they lose their charcoal grey tiles to show the concrete they are made of.

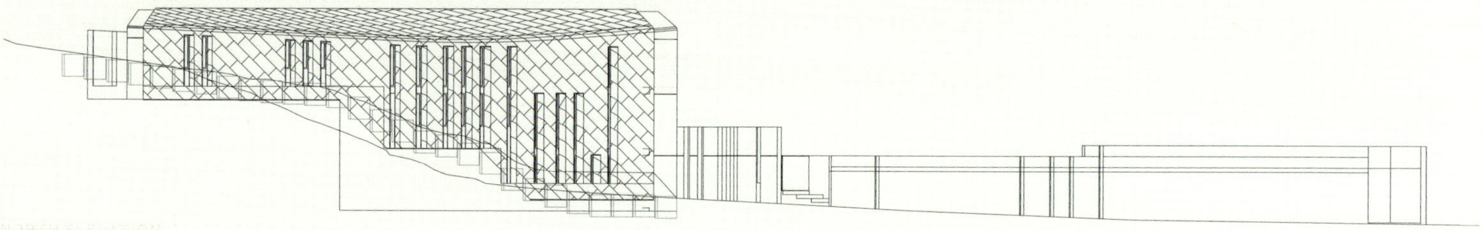

NORTH ELEVATION

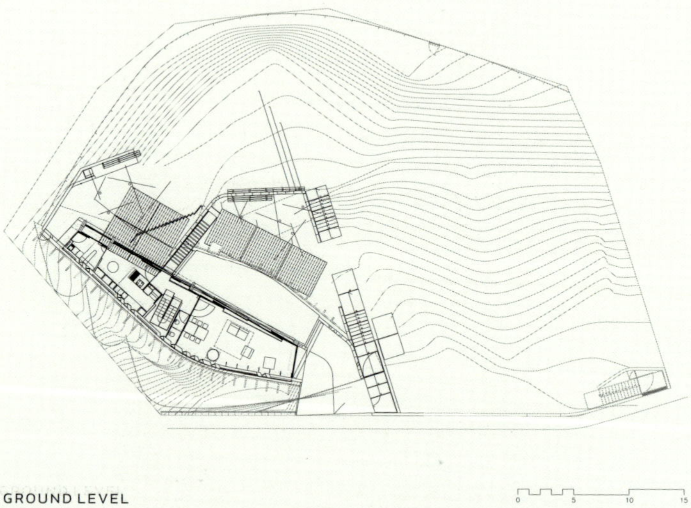

GROUND LEVEL

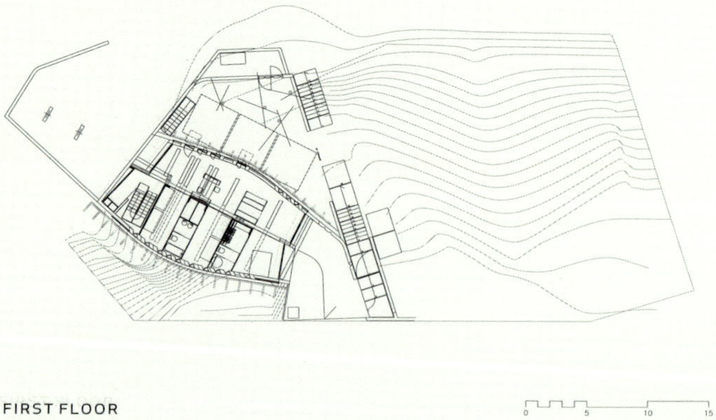

FIRST FLOOR

ARCHITECTS: Equip XCL. Xavier Claramunt, Candi Casadevall / **LOCATION:** Santa Margarida de Montbui, Barcelona province / **CHRONOLOGY:** 1992 – 2004 / **COLLABORATORS:** Martín Ezquerro, Yago Haro, Marc Zaballa, Miquel de Mas, Pau Vidal, Xavi Medina / **CLIENT:** Jordi Muntaner y Elena Ramón / **SURVEYOR:** Joel Vives / **BUILDER:** Construccions Joan Ribas / **BUILT SURFACE:** 400 m^2 / **TOTAL COST:** 450.000 € / **PHOTOGRAPHER:** Eugeni Pons

CERAMIC: Exterior stoneware tiling over tongue and groove ceramic backing attached to galvanised steel understructure. Matte black 5 x 5 cm Gresite mosaic tile

José Durán Fernández

Calle San Vicente

Burriana, Spain

Tile of Spain Architecture and Interior Design Awards 2007

The gradual deterioration of Calle San Vicente, its undeniable historical value and its importance as the main artery through the centre of Burriana were the main reasons behind the designers of this project deciding to act boldly on paving and lighting and to enhance its role as a pedestrian thoroughfare. The design adopted responded to a complete regeneration of its layout, on both visual and ambience levels. Ceramic was chosen as the only paving material base, given its significant historical background. An in-depth study was made of the inherent qualities of the material and its new application. In this way, a new tile with new dimensions and shape was designed. A ceramic cobblestone was developed.

In contrast to a mostly grey town, the project created a colourful, almost modernist, design for the route leading from the main square of Burriana (El Pla) to the Carmelite Church. The street was laid out in 6 sections - 6 experiences, which could be extrapolated to the whole town by means of coloured bubbles.

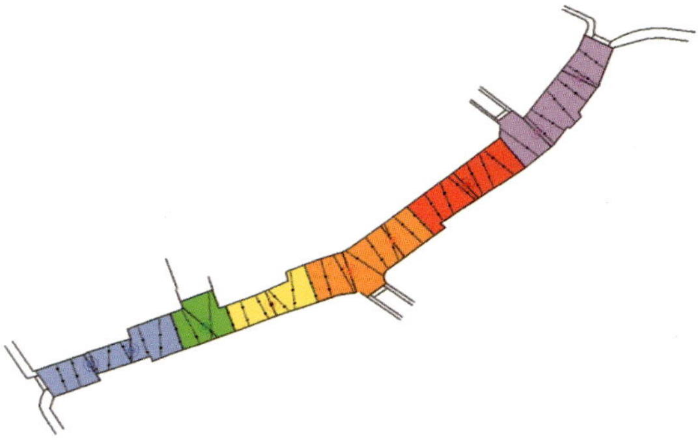

SELECTION OF COLOURS USED

ARCHITECT: José Durán Fernández / **LOCATION:** C. San Vicente, Burriana, Castellón Province / **PROJECT:** Feb 2006 – Sep 2006/ **CONSTRUCTION:** Sept Sep 2006 – Sep 2007 / **COLLABORATORS:** Aleixandre Franch (Civil Engineer), Alicer (Javier Mira Peidró), Javier Cebrián Enrique (Architect), Víctor fe Alba (Graphic Design), Dicolaser,s.l. (Discs) / **CLIENT:** Burriana City Council / **LIGHTING:** Iguzzini / **BUILDER:** Lubasa / **BUILT AREA:** 1,301 m^2 / **TOTAL COST:** 410.146,06 € / **PHOTOGRAPHERS:** Vicente fnd Alejandro Peris

CERAMIC: Stoneware ceramic cobblestone 100 x 100 x 20 mm striated upper surface and glazed in lower surface with box joint

Equip Arquitectura
Pich-Aguilera
Felipe Pich-Aguilera,
Teresa Batlle

Telefónica Móviles Call Center

Toledo, Spain

The building was conceived as a large container where different industrially-produced parts have been assembled. The design, materials and details of construction incorporate systems that take advantage of the environmental conditions and support the temperature control of all of its spaces. Two, three-dimensional core structures and a portico support huge concrete slabs, resembling a multi-level bridge. These are the large platforms housing the call centre work areas.
The central area of the building is an open space crossed by 5-metre wide walkways supported by core areas containing services and communications. The idea was to create a space with great plasticity and flexibility, with the purpose of allowing every individual work space and areas of busy activity to benefit from the volume of the building.
The design sought to bring the shape of the building together with its bioclimatic function and to create an industrial system of construction that would determine the resulting volume and layout of the building.

The building is clearly divided into three distinct functional areas:
The first area houses the operator service on all levels. This area works like a large climate-controlled atrium. The roof covers the entire space and, by means of different angles, layers and mechanisms, provides maximum heat transfer and distribution of air (hot or cold) to all areas. All natural light in the building enters through two louvred, tiled curtain walls, especially designed to control the sunlight entering the building and allowing enough light to reach all work areas.
A second area, closely joined to the first, contains the rest area. It works like a large greenhouse that contributes to the comfort of the other spaces in winter and is three-tiered, giving it direct access to each of the work platforms.
The third area is where offices, control, classrooms and parallel services are located. It is laid out as a set of enclosed spaces and half-covered courtyards. In this area, the raised walkways come together and are enclosed to form compartmentalised work areas overlooking covered courtyards at every level.
This north-facing area, with its green courtyards works as a temperature difference regulator. Vegetation plays a vital role owing to its natural cooling effect and sound-absorbing qualities. The vegetation is sustained thanks to roof rainwater collection tanks on the terraces and where they are cantilevered over the portico. The roof tanks collect and store the rainwater that serves as hydroponic substrate for the vegetation.
The main façades of the building are curtain walls clad with large tiles.
The collaboration of the Clay Technology Centre of the Castilla-La Mancha Autonomous Region was fundamental in supporting the involvement of all of the area's tile manufacturers in the project and facilitated the viability of a locally-produced solution for the façade so that it could be adapted to factory processes and systems. The company awarded the tender was responsible for supplying and installing the material.
The façade is made up of large ceramic tiles.
The integral concept of a double-skin curtain wall (polycarbonate panels on the inside and tiles on the outside) guarantees continuous ventilation, an evacuating convection effect – eliminating possible transfer of heat or cold from outside and watertightness, and improves acoustic absorption properties of the outer skin.

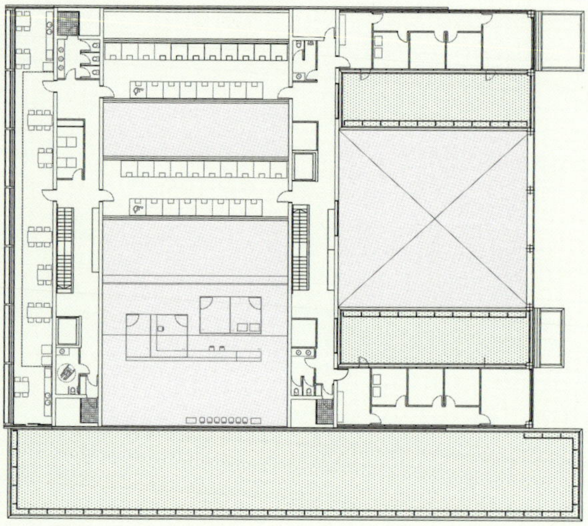

FLOOR PLAN LEVEL THREE

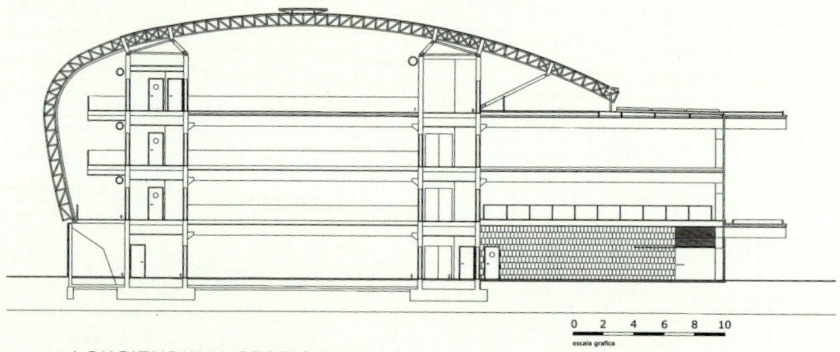

LONGITUDINAL SECTION

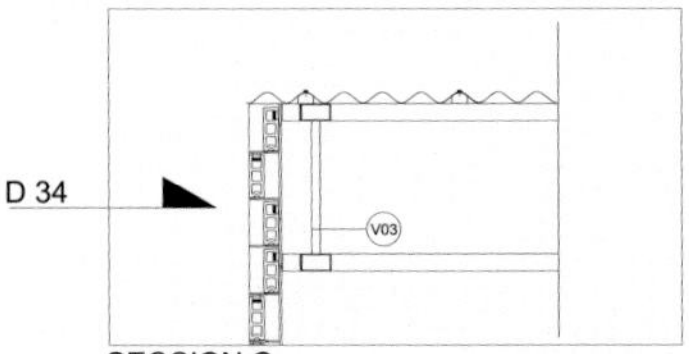

SECCION C

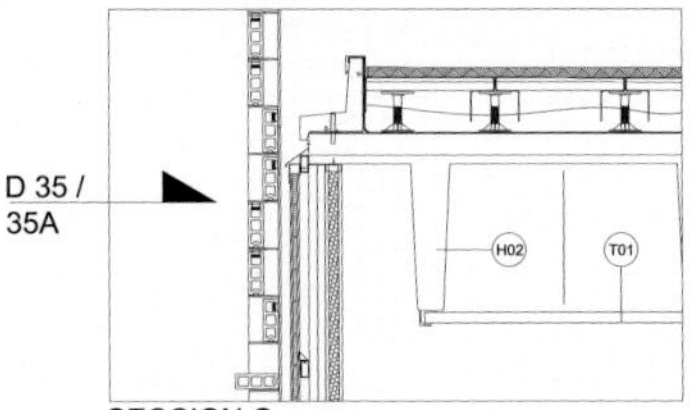

SECCION C

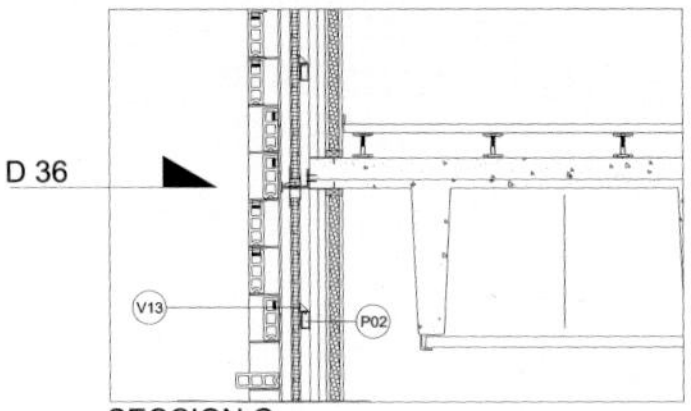

SECCION C

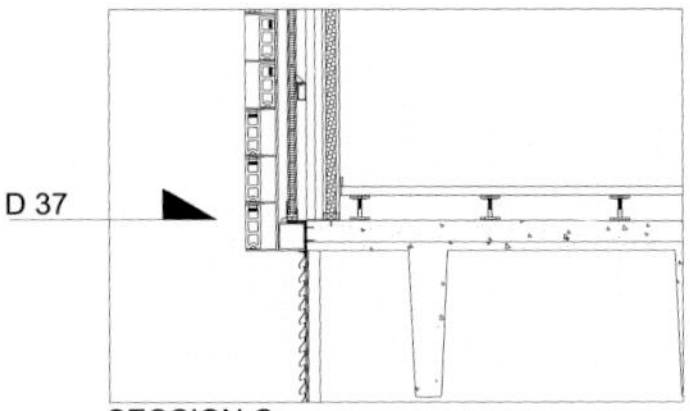

SECCION C

0 50 cm

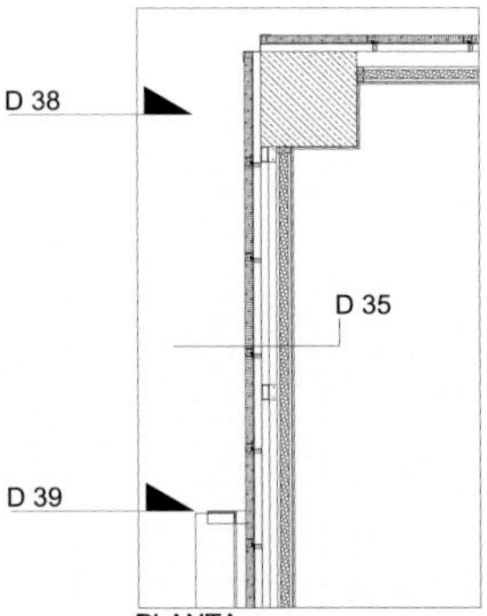

PLANTA

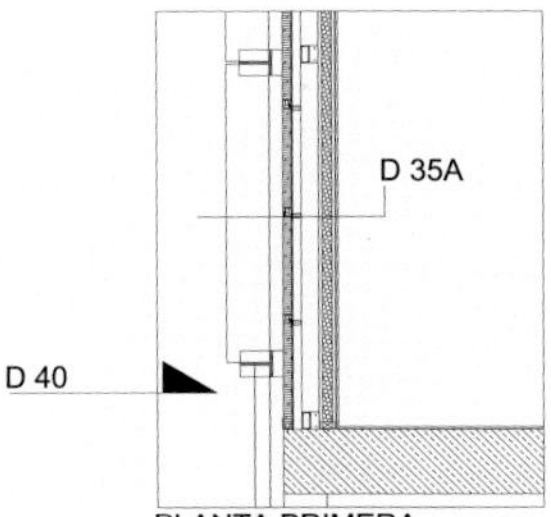

PLANTA PRIMERA

0 50 cm

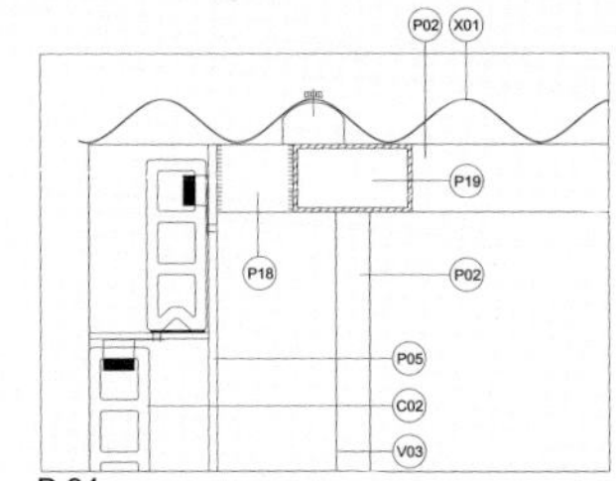

D 34

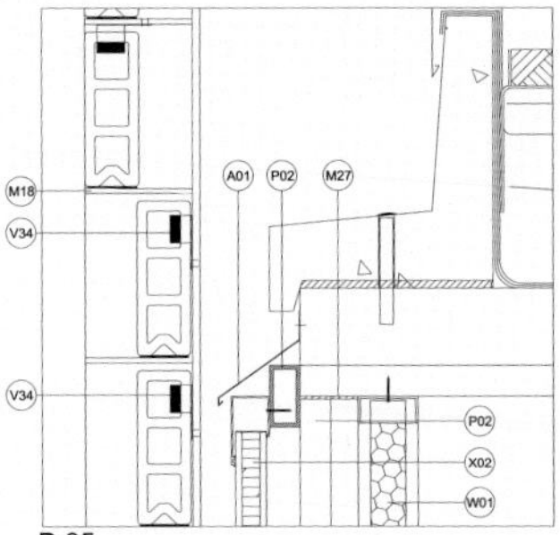

D 35

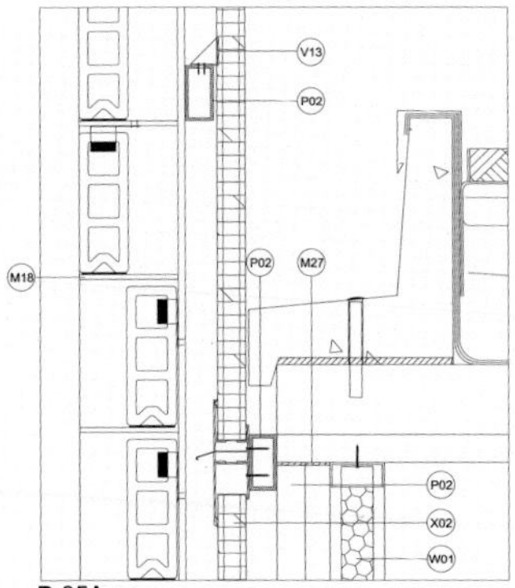

D 35A

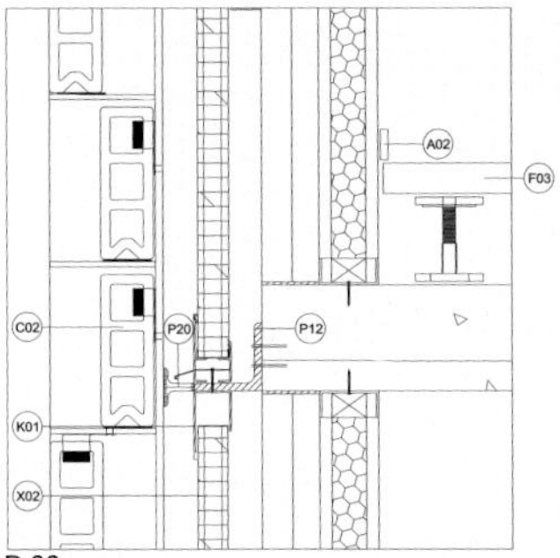

D 36

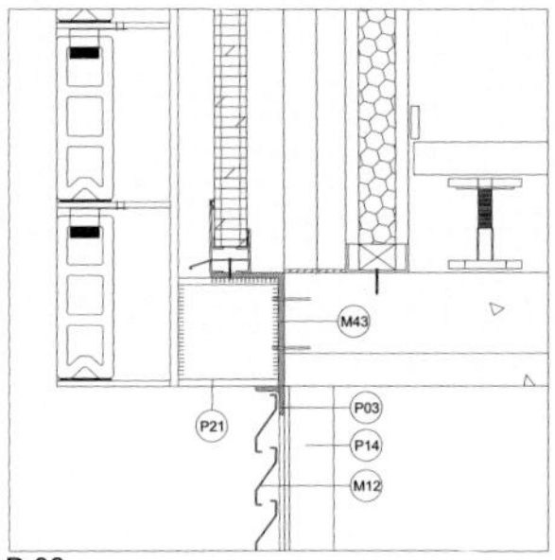

D 06

0 20 cm

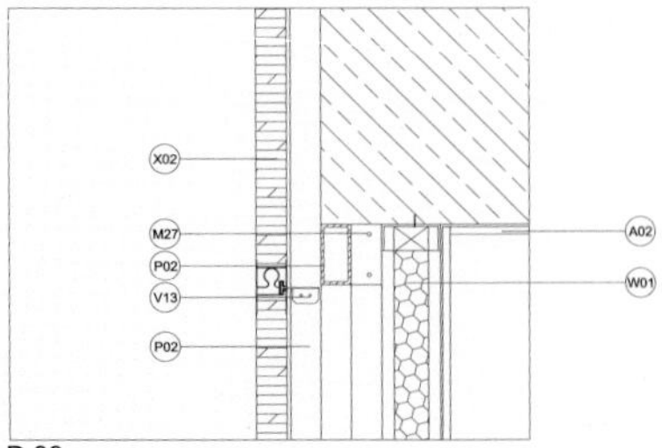

D 38

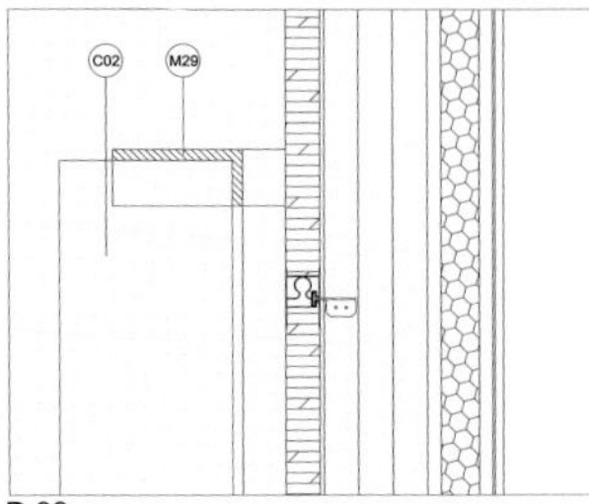

D 39

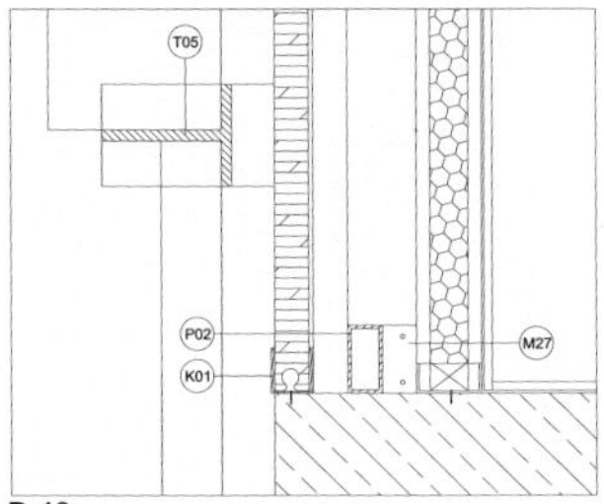

D 40

PLANTA BAJA

MATERIAL'S LIST

M12 'Z' shaped profile, Hiansa PE-570 or similar, hammered painted
M18 140x70x7 plate, hammered painted
M27 75x75x5 anchorage plate, hammered painted
M29 140x60x10 'L' shaped profile, hammered painted
M42 (180+50) x5 plate
M43 (140+90) x200x7 painted plate
A01 Aluminium top, folded metal sheet, anodized, silver colour
A02 40x10 aluminium pipe as skirting board, silver colour, anodized
T01 Aluminium, 30 cms. strips, guide
H01 'L' shaped runoff, dealt with automatically, joined to the slab with mechanical fixation
C01 14 cm x 28 cm vitrified plain tile, various colours from Covallay or similar
C02 Big size piece 20x70x140 cm Bardeau de Guiraud Frères/ St. Gobain or similar
I02 Thermal insulated plate made of extruded polystyrene (4 cms)
P02 40x80x4 tube
P03 L 35x35x6
P05 T 140
P12 L 90x90x10
P14 T 50x50
P18 80x80x4 tube
P19 140x80x5 tube
P20 Cutted T50
P21 HEA 140
X02 Lexan Thermoclick Sheet system LTC40/4RS3600 plate from GET or similar.
W01 Partition, hidden profile, double 16/450/16, metallic fittings, thermal insulation, melamined wooden panels from Supra or Desmon or similar.
V02 Twisted steel cable 3 mm. in diameter.
V03 Covered truss
V06 40x40 wooden batten
V13 Anchorage for the Xo2
V34 Mechanical anchorage with neoprene for CO_2
K01 Standard profile for polycarbonate (see. Installer)
K02 Aluminium frames, anodized 25 micron, silver colour

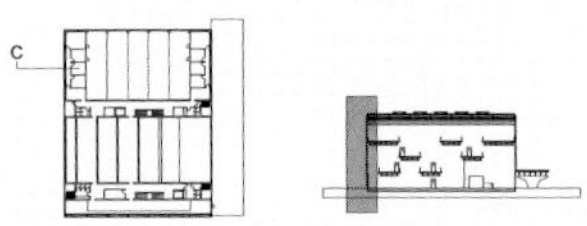

CERAMIC SKIN. DETAILS OF CONSTRUCTION FROM VERTICAL AND HORIZONTAL SECTIONS OF FAÇADE

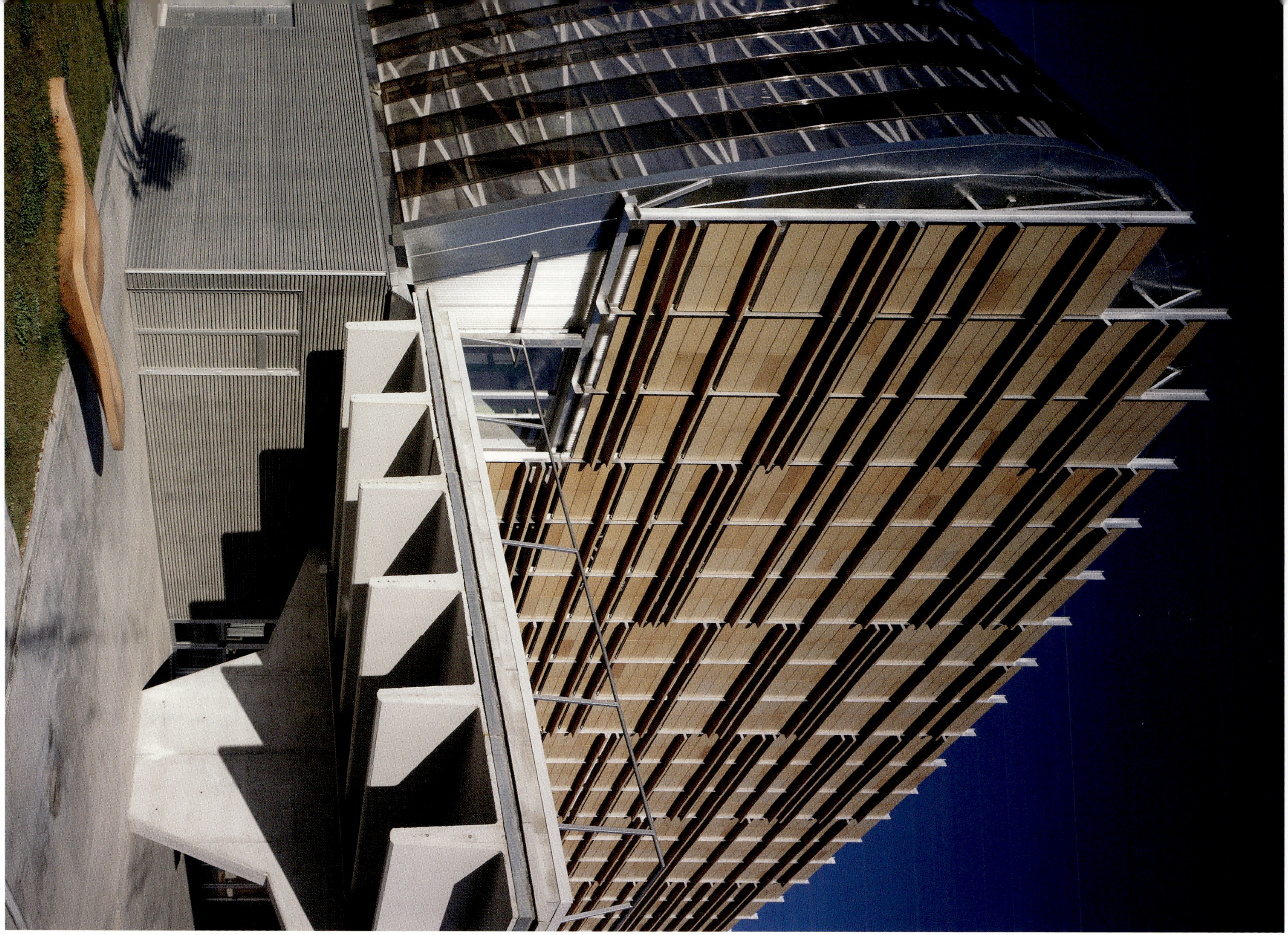

ARCHITECTS: Equip Arquitectura. Felipe Pich-Aguilera. Teresa Batlle / **LOCATION:** Centro de Instalaciones II, Toledo / **COMPETITION:** 2001 / **PROJECT:** April 2001 – September 2001 / **CONSTRUCTION:** August 2004 – August 2005 / **JEFES DE GRUPO:** Xavi Milanés, Ángel Sendarrubias / **PROJECT MANAGERS:** Bruno Sauer, Ute Müncheberg / **COLLABORATORS:** Ivan Acevedo, Pepe Vázquez / **CLIENT:** department of industry, government of Castilla-La Mancha Gicaman s.a. / **STRUCTURE:** BD, Consultors / **SYSTEMS:** PGI Grup / **SURVEYING AND ESTIMATES:** Tècnics G3 / **RENEWABLE ENERGIES:** Emilio Miguel Mitre. EMMA, Consultores / **BUILDER:** Construcciones Ortiz / **PHOTOGRAPHER:** Eduardo Sánchez López

CERAMIC: Louvred curtain wall. Large ceramic tiles (60 x 30 x 2,5 cm) stoneware integrated curtain wall system glazed ceramic tiles (14 x 28 cm) two colours

Tomás Alía

Ceramic concert

Espacio ASCER Casa Decor Madrid '07

Madrid, Spain

Space and ceramic were the fundamental and unifying elements in the development of this project. A sculpted shape was designed to represent a tower reaching to the sky and to receive light from above. The sculpture stands at almost six and a half metres tall and was conceived to take advantage of the height of the courtyard without smothering it, and to make the ceramic appear imposing while highlighting the great skill involved.
The sculptures that define the space are a series of shapes laid out to allow the viewer to enter and pass through them and at the same time touch and appreciate the material.
The space comprises six free-standing elements, another two against side walls and a sculpted seat at the entrance.
Each of the elements has a different shape, size and design with a series of cuts through which the viewer can pass.
The pieces are completely covered in 14 different ceramic tiles and laid out to give different textures according to the different parts of the sculptures. They add variety and force to the group.
There are two types of lighting systems: uplighting serves to enhance the features of the material while lights in the cut out areas change colours to give a sense of energy to the space.
The floor is covered in black carpet, except for the access ways that lead viewers to walk inside the sculpture. These are made of tiles alternating with black carpet cuttings.

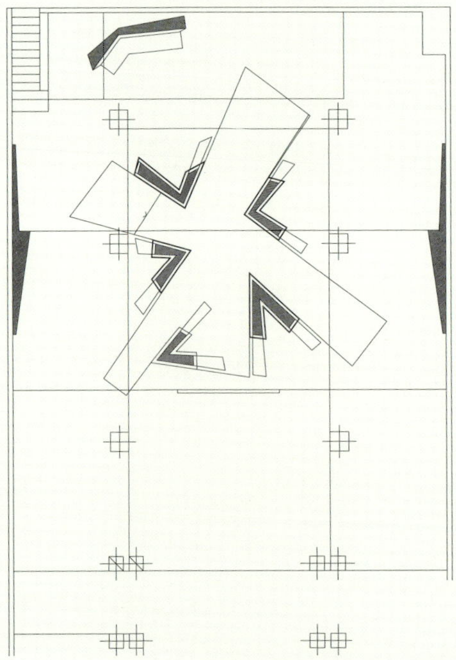

PLAN

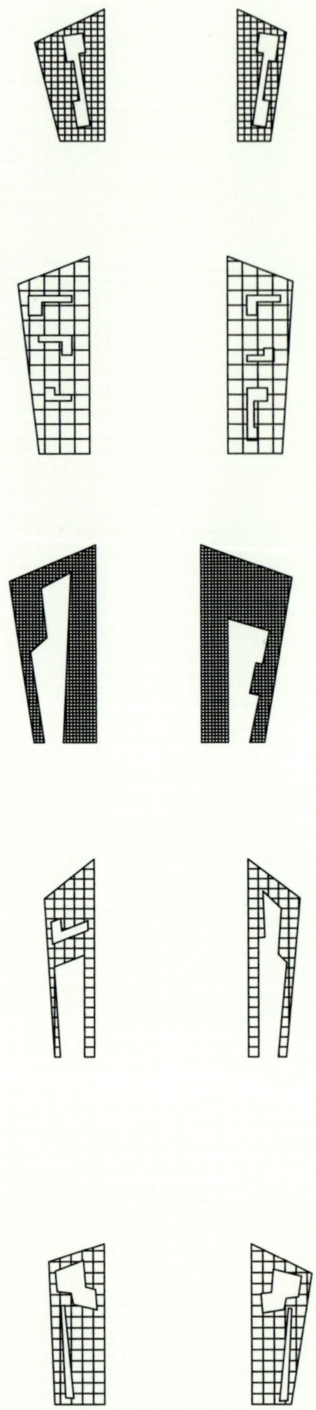

FREE-STANDING FAN-SHAPED CERAMIC SCULPTURES

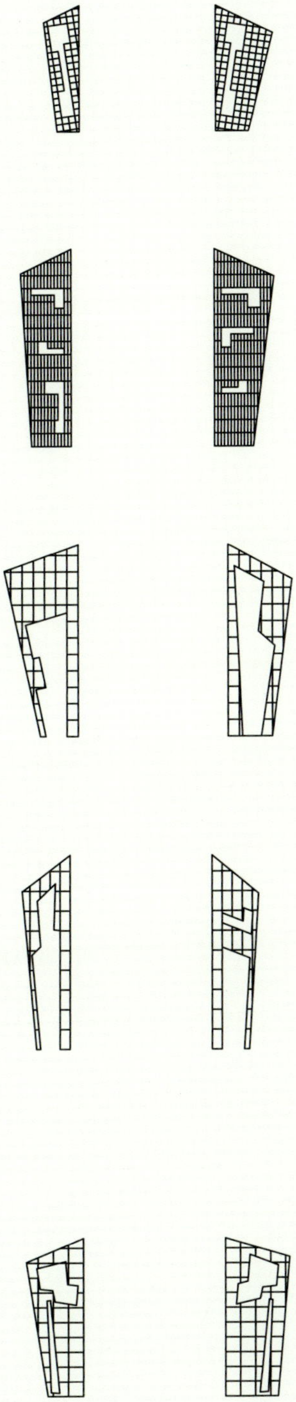

ARCHITECT: Tomás Alía / **LOCATION:** Calle Alcalá 8, Madrid / **CHRONOLOGY:** 2007 / **CLIENT:** Ascer / **BUILDER:** Efficio / **LIGHTING:** Sadilsa / **BUILT AREA:** 220 m²

CERAMIC: Stoneware and porcelain stoneware wall and paving tiles of different sizes

Archea Associati
Laura Andreini,
Marco Casamonti,
Silvia Fabi, Massimiliano
Giberti, Gianna Parisse,
Giovanni Polazzi

New municipal library

Nembro, Italy

This building dates back to 1897 and was designed as a primary school. It changed its use a number of times – town hall and nursery school, before finally becoming a library. The local council decided to rescue the building from its then state of neglect and rehabilitate it as a library. The strategic location of the building inside the town, the U-shaped design of the original building and the need for extra space led to the project for increasing the total building area by adding a new wing to close the open side and create a courtyard.
The entrance to the old school was retained for the library and opens into an area used for reception and information. The design modified and opened up the existing rooms to achieve an airy, emblematic space.
Once past this area, the glassed-in portico is reached, which connects the different areas of the library.

On the ground floor, a computer room for data base searches and consultation opens onto a new space that leads to the reading room. This level and the first floor of the old building house reading rooms, the periodicals library and a multi-purpose room while the new block houses the book collection and a reading room. This new area, completely transparent, is characterised by its skin comprising 36 x 36 cm red-glazed terracotta tiles over a framework of steel profiles; this is a curtain wall capable of subduing and filtering the natural sunlight, separating the interior from the exterior. The tiles are freely positioned alternating light and shade. The material chosen, while evoking past tradition, also has a modern aspect in its specific function and in the way it is attached. Its physical and expressive nature demonstrates a capacity to endure in a context with few references.

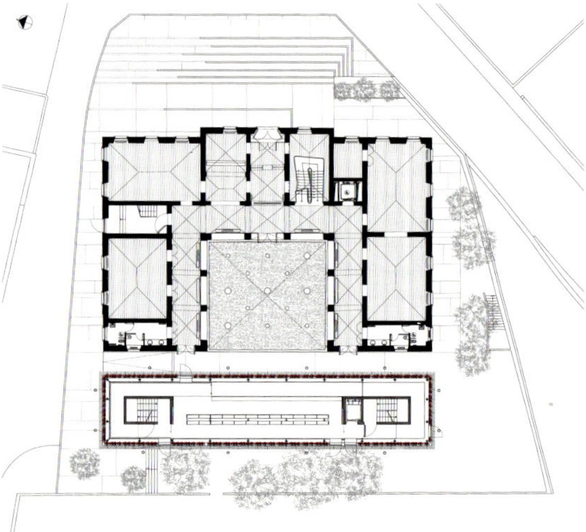

GROUND LEVEL

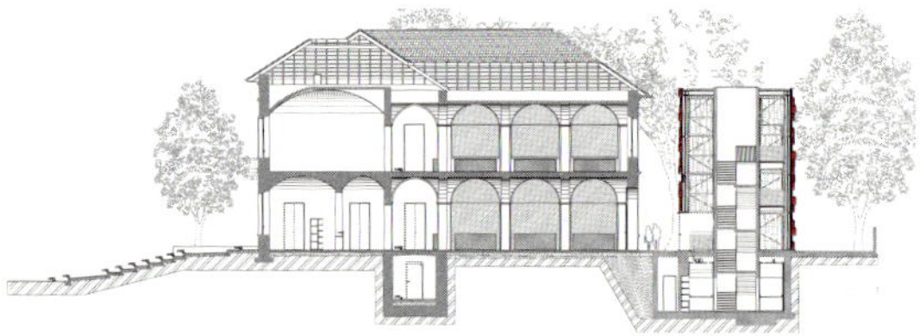

TRANSVERSAL SECTION

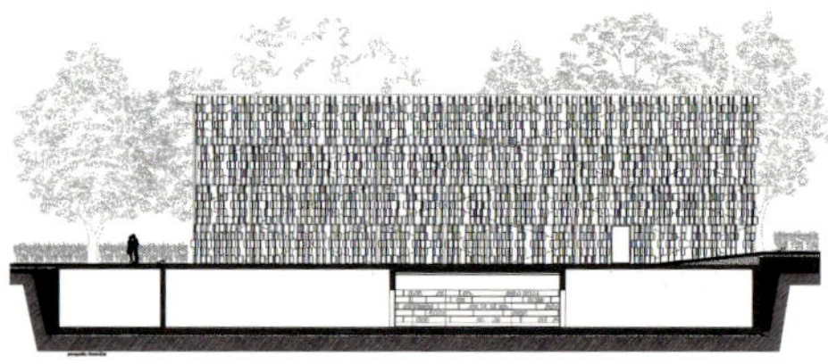

ELEVATION EXTENSION

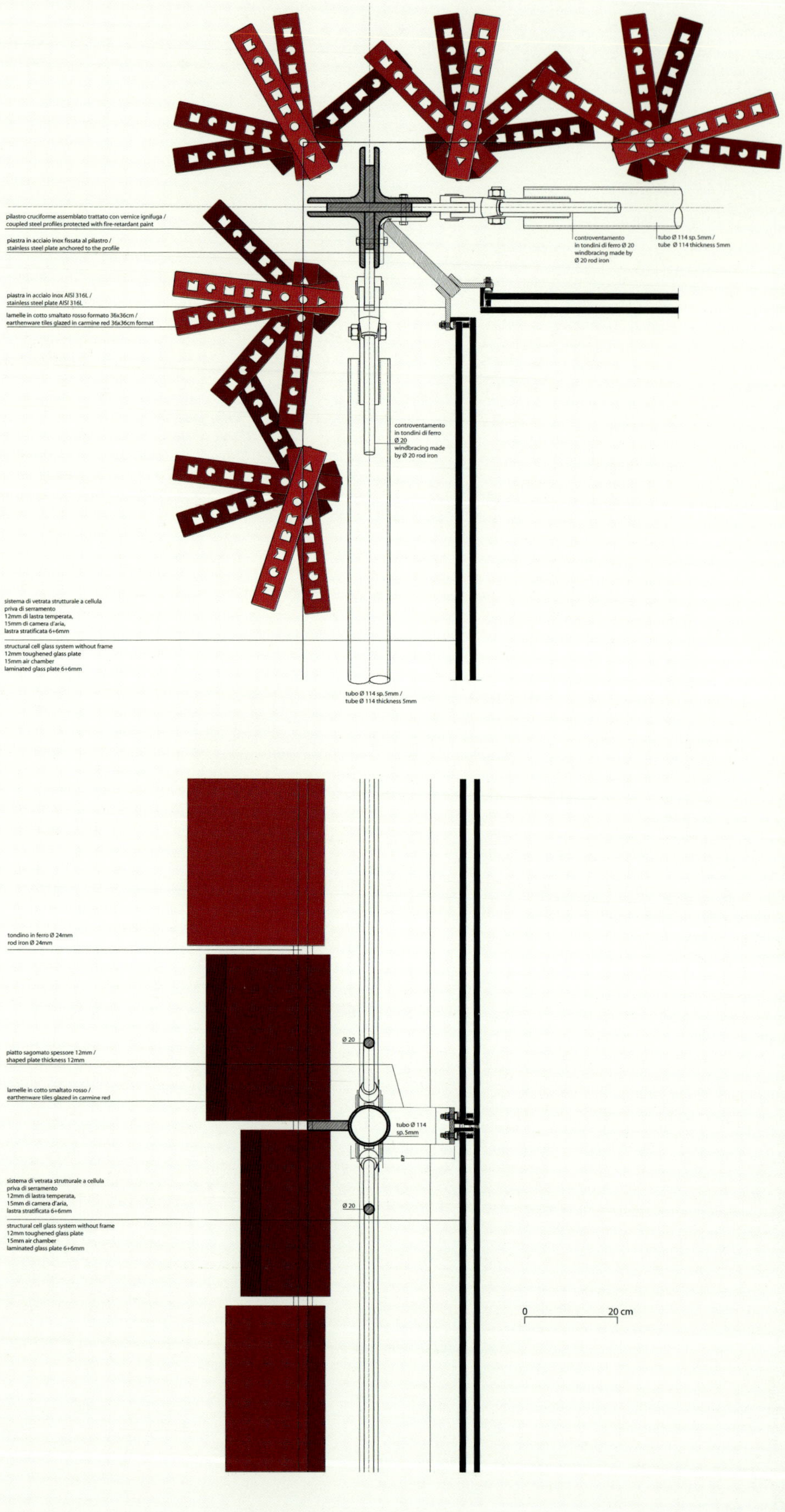

CURTAIN WALL. CORNER DETAIL

ARCHITECTS: Archea Associati. Laura Andreini, Marco Casamonti, Silvia Fabi, Massimiliano Giberti, Gianna Parisse, Giovanni Polazzi / **LOCATION:** Nembro (Bergamo), Italia / **PROJECT:** 2000 – 2001 / **CONSTRUCTION:** 2002 – 2007 / **COLLABORATORS:** Francesco Giordani, Giuseppe Pezzano / **CLIENT:** Comune di Curno / **STRUCTURE:** Favero e Milan Ingegneria / **LANDSCAPING:** Studio Archea / **INSTALLATIONS:** Favero e Milan Ingegneria / **ELECTRIC INSTALLATIONS:** Favero e Milan / **LIGHTING:** Favero e Milan Ingegneria / **BUILDER:** Zeral srl / **BUILT AREA:** 1.800 m² / **BUILT VOLUME:** 11.200 m³ / **TOTAL COST:** 1.900.000 €

CERAMICS: Outer curtain allowing for daylight control. It is made up of 36 x 36 cm clay pieces enamelled in red and mounted on a steel frame

34

Javier Bernalte, José Luis León

Spa

Ciudad Real, Spain

This project for a spa focuses on human awakening. It recreates the maternal womb that encloses warm fluids with a continuous skin through which diffused light penetrates to create a state of total peace.
The concept was therefore to install a continuous, organic skin in an empty space on the ground floor of a residential block.
A project calling for the appearance either of dry surfaces on walls and ceilings or of moist skin in the bath area provides the perfect opportunity to give maximum visibility to ceramic, in this case white vitrified mosaic tiles.
Weightless, within this glowing skin, a stone dihedron brings us back to reality. It represents a useful space floating over the liquid element. The linear commercial area is laid out over the horizontal plane of the dihedron with products displayed in successive cut outs in the vertical plane. Light flows out of the bays after being reflected in the vitreous skin surrounding the dihedron. The vertical plane also appears weightless as light flows out through its edges.
The sales area storage space is hidden in the lower part of the vertical plane behind pivoting stones set into the wall. At the back of the sculpted dihedron, the stone shows through the light by means of a carved alabaster slab with the logo. Next to the stone dihedron, a linear wooden unit parallel to the vertical facing is for service use and also serves to separate the bath area.
The elements making up the project are completed by a sheet metal structure housing the furnace and boiler, immersed in the thermal bath, and by a lit glass structure housing the sauna.

Termal ORETANIA
Horario
14:00

1 ENTRADA
2 TIENDA
3 ATENCION AL PUBLICO
4 ASEO MASCULINO
5 ASEO FEMENINO
6 VESTUARIO MASCULINO
7 VESTUARIO FEMENINO
8 ALMACENAJE
9 SAUNA
10 HOGAR
11 TERMA
12 TERMA DINAMICA
13 AREA DE MASAJES

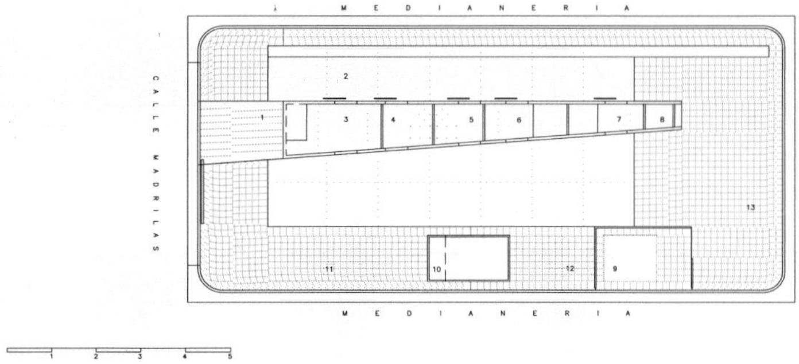

GROUND FLOOR

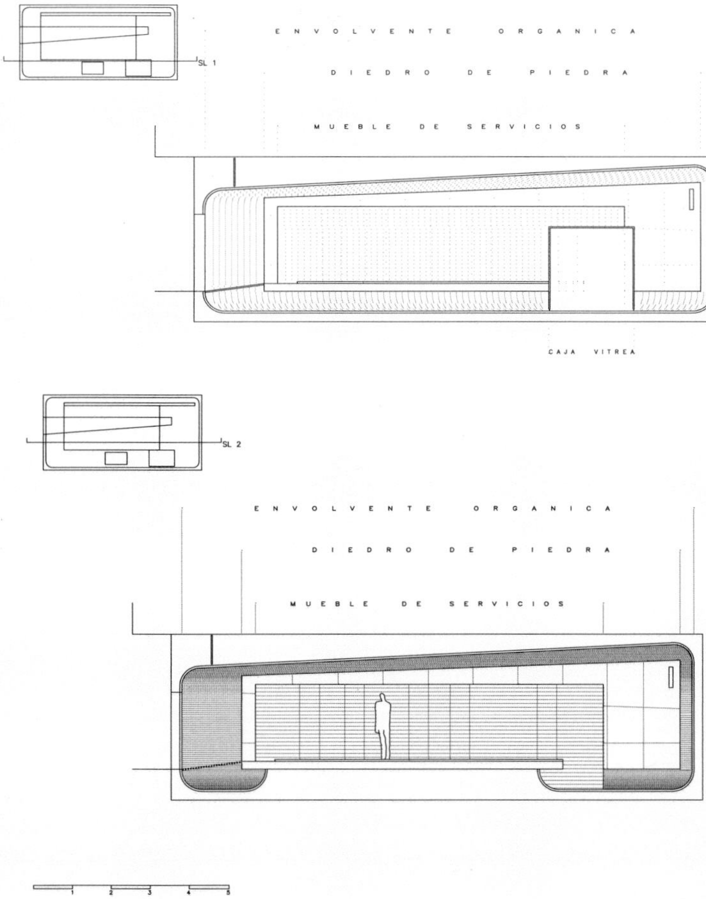

LONGITUDINAL SECTIONS

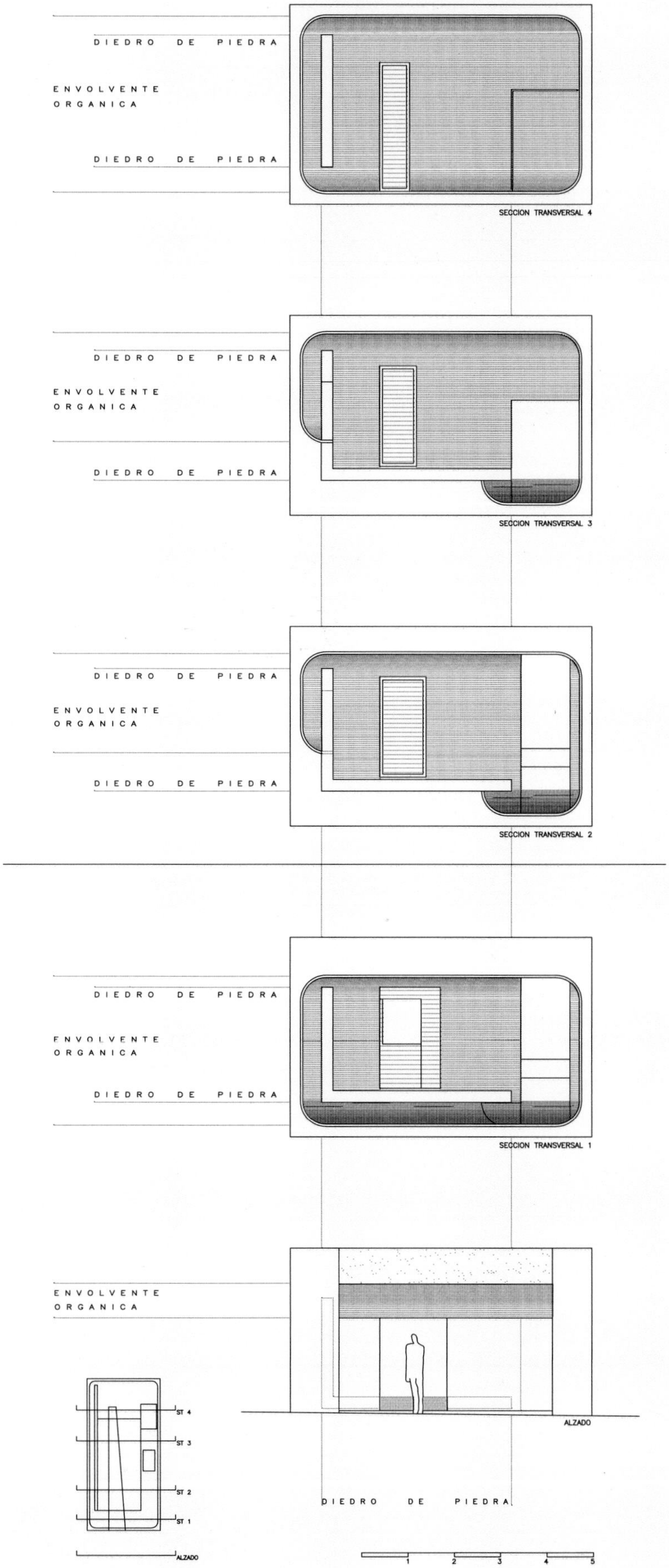

TRANSVERSAL SECTIONS

ARCHITECTS: Javier Bernalte, José Luis León / **LOCATION:** Ciudad Real, Spain / **CHRONOLOGY:** March 2001 – August 2001 / **CLIENT:** Soler y Cuesta S.L. / **COLLABORATOR:** Marisa Porras / **PROJECT MANAGERS:** José Luis León, Javier Bernalte / **BUILDER:** Construcciones Castillo de Daimiel S.L. / **PHOTOGRAPHER:** Angel Baltanás

CERAMIC: White vitreous mosaic tiles 20 x 20 mm attached to floor, ceiling and walls with white waterproof white cementitious-based tile adhesive

Alberto Burgos

Patio House

San Javier, Spain

A dwelling inside a patio is a common type of dwelling in the Murcia region. Three sides are enclosed by walls and the fourth side has a swimming pool.
The design is for a dwelling surrounded by walls that close it from view on three sides. A perpendicular swimming pool at lower floor level, closes access to the open end without blocking views over a public park.
A roof part of the lower area and under this, two parallel walls break up the space lengthways. Between them, light partitions form rooms. Adjoining the outer walls, two differentiated outbuildings serve as utility areas - garage and laundry.
Materials are important. For this design, terracotta tiles are essential, guaranteeing continuity between interior and exterior regardless of climatic conditions. It is warm in winter and cool in summer. Pleasant to touch, you can walk barefoot on it from inside to outside, from the shade of the dining room to the sun shining over the pool.

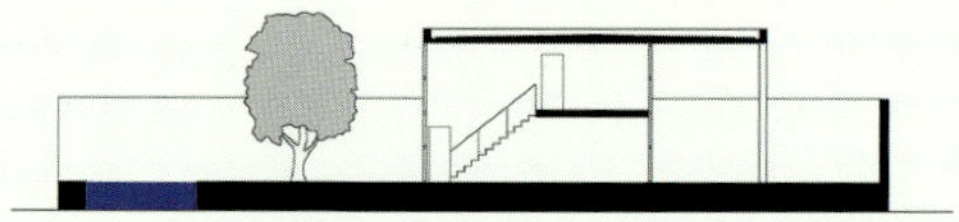

LONGITUDINAL SECTION

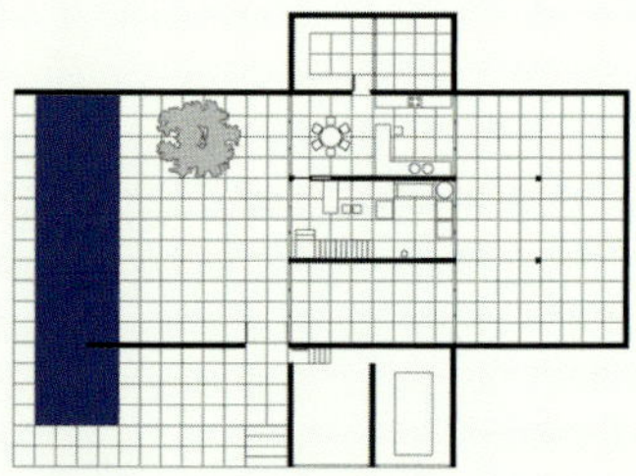

GROUND FLOOR

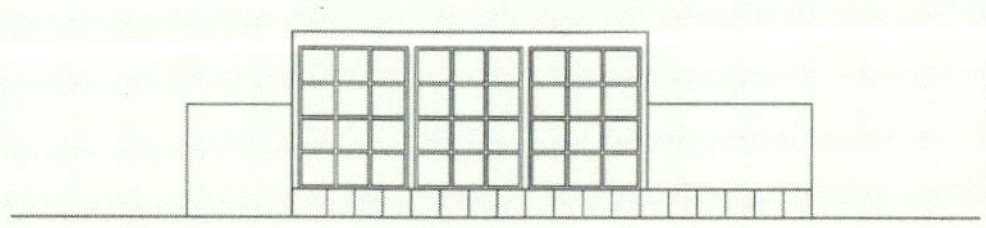

MAIN ELEVATION

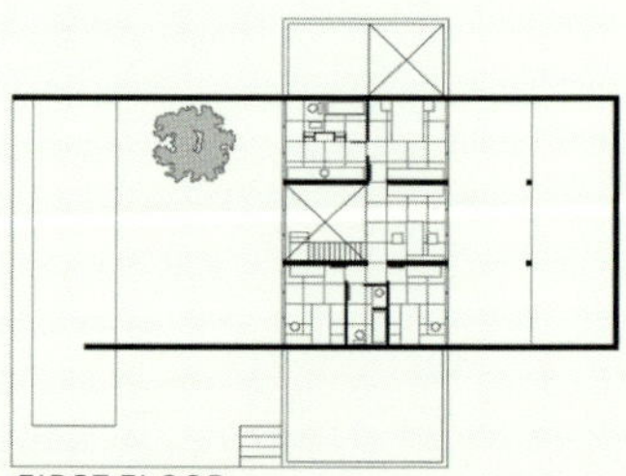

FIRST FLOOR

ARCHITECT : Alberto Burgos / **LOCATION :** San Javier, Murcia / **PROJECT :** 1990 / **CONSTRUCTION :** 2007 / **COLLABORATORS :** Javier Rubio / **CLIENT :** José Antonio Bastida / **STRUCTURE :** Adir s.l. / **BUILDER :** Miguel Mengual (Adir s.l.) / **QUANTITY SURVEYOR :** Joaquín Martínez / **BUILT AREA :** 296 m² / **PHOTOGRAPHER :** Salvi Vivancos

CERAMIC : 30 x 30 cm semihand-made tiles were used for the entire lower level, inside-outside, horizontal-vertical 50 x 50 cm tiles were laid adjusted to the 400 x 400 grid of the house the pool is faced with ultramar 115 x 240 cm glazed tiles the bathrooms are faced with 20 x 20 cm white stoneware tiles

40

EMBT Arquitectes
Benedetta Tagliabue

Hafencity public spaces

Hamburg, Germany

This project is the result of an international design competition organised to promote the urban renewal of the Elbe riverbanks in the centre of the city of Hamburg. The square at Sandtorhafenkopf is the first stage of the Hafencity redevelopment project.
It is the connection between the traditional-style port basin and the Sandtorpark, starting point for all of the promenades that follow the lines of buildings at the edge of the basin.
The square features wide stairways and ramps descending in different levels towards the water. The direction of the stairs varies, resulting in different spaces and rest areas. Benches define the routes and offer views of the water and the boats moored in the basin.
Cafés with outdoor seating provide the square with a lively atmosphere.
There are two basic lighting systems: a linear system following paths and promenades, and specific lighting for different settings and squares.

The space is designed with different materials and colours. Asphalt is used for pavements on Kibbelstegstrasse, with a scattering of tree wells covered with grass or gravel. The use of prefabricated concrete on stairs and ramps alternates with stone paving, masses of prefabricated concrete bird shapes and coloured mosaics made from terracotta tiles forming abstract fish designs. Vertical surfaces are also encrusted with coloured bricks, slightly protruding – depending on the pattern – creating stylised shapes of marine creatures.
Enormous blocks of prefabricated concrete, reminiscent of port structures are accompanied by free-style railings to define the water's edge. The significant difference in tidal levels are marked in the prefabricated elements lining the wall that separates land from water.

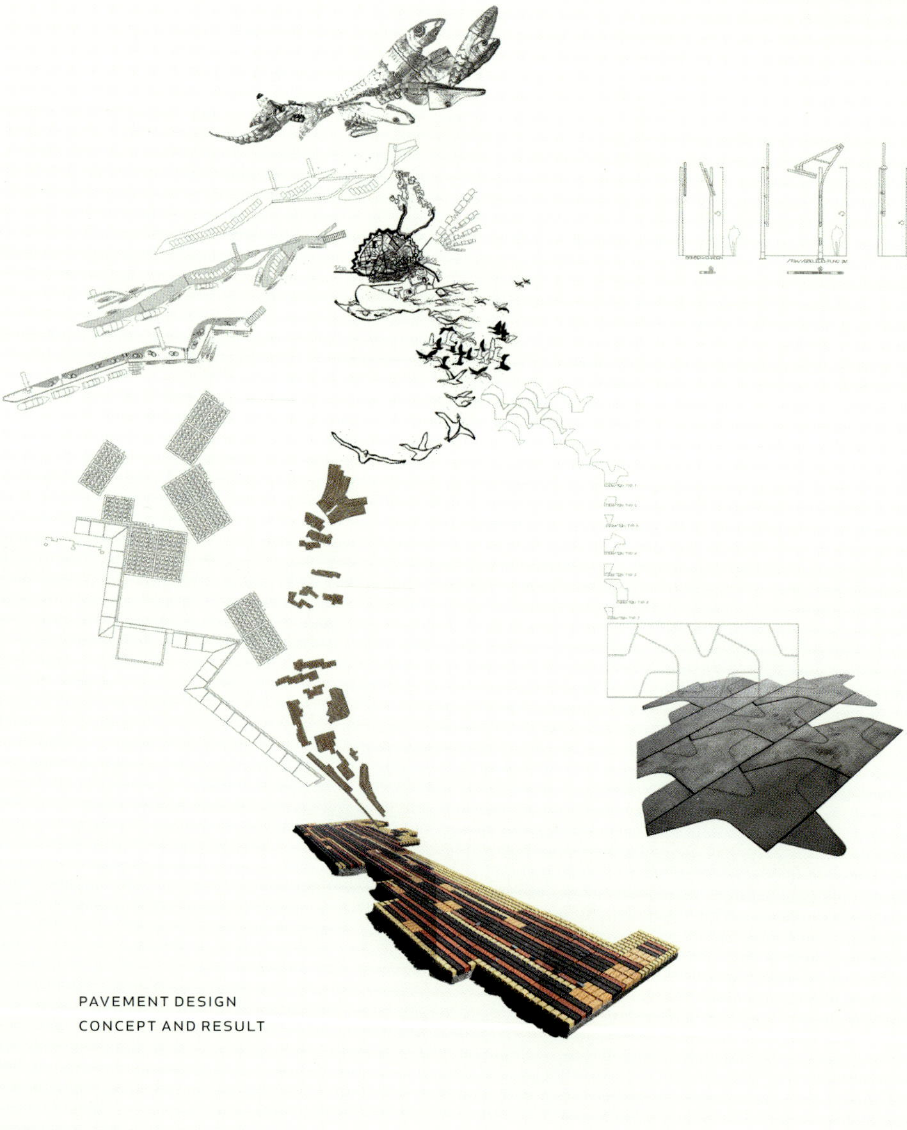

PAVEMENT DESIGN
CONCEPT AND RESULT

LIGHTING SYSTEM
PROFILES

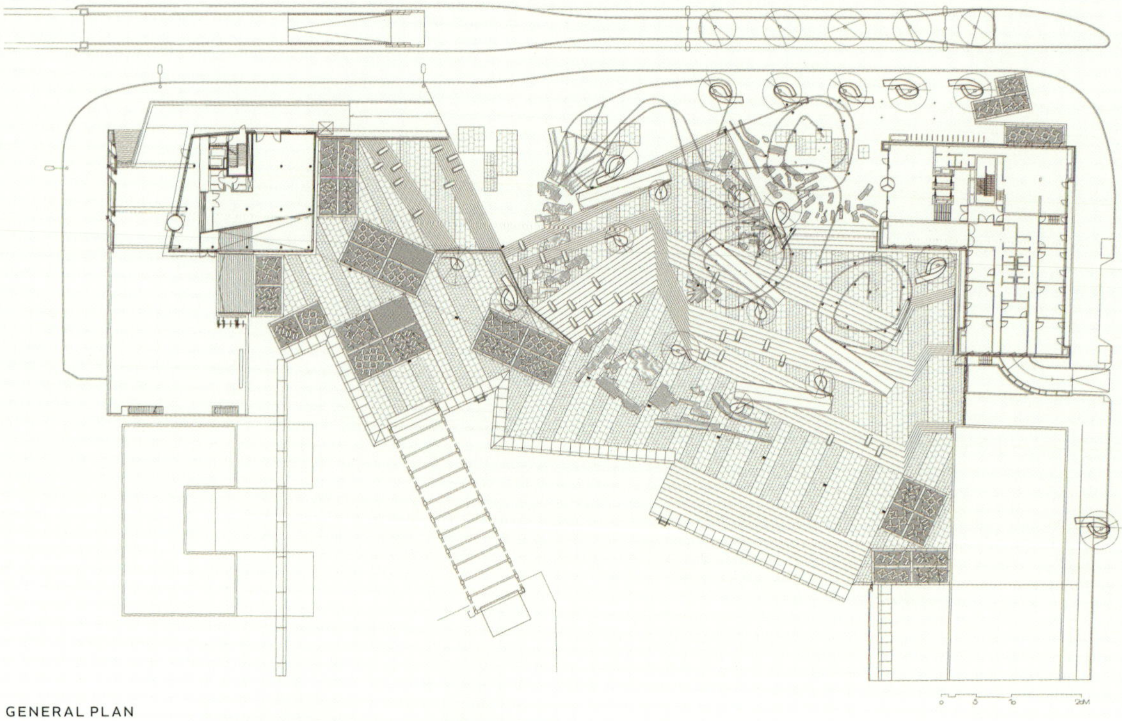

GENERAL PLAN

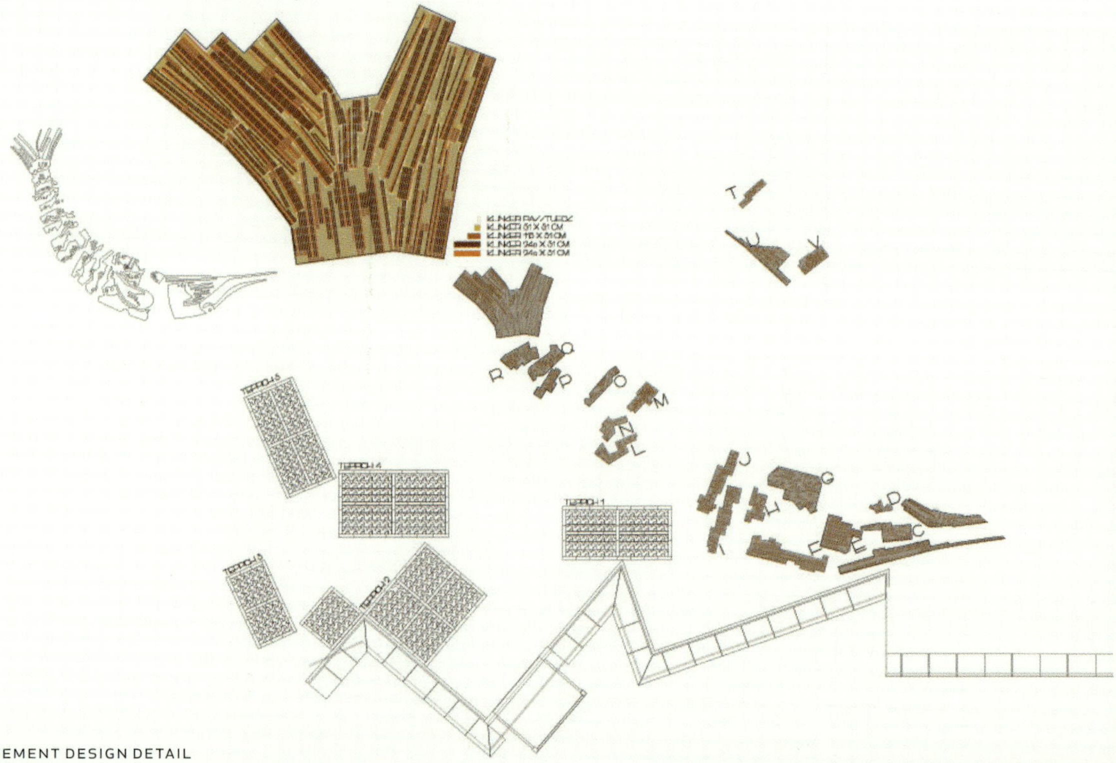

PAVEMENT DESIGN DETAIL

ARCHITECT: EMBT Arquitectes. Benedetta Tagliabue / **LOCATION:** Hamburg, Germany / **COMPETITION:** 2002 / **PROJECT:** 2003 – 2006 / **CONSTRUCTION:** 2006 – 2007 / **ASSOCIATED ARCHITECTS:** Karl Unglaub (project manager) Stefan Geenen, Elena Nedelcu / **COLLABORATORS:** Elena Rocchi, Jorge Carvajal, Eugenio Cirulli, Massimo Chizzola, Santiago Crespi, Marco de Gregorio, Annunziata Dezio, Daniel Domingo, Jörgen Dreher, Gianfranco Grondona, Joana Guerra, Simon Junge, Irene Kasjanenko, Maximilian Kneucker, Jan Kokol, Ana Catarina Miguel, Beatriz Minguez, Amanda Mori, Gordon Moss, Fernando Mota, Lucía Ortiz, Adelaide Pasetti, Jakob Pitroff, Judit Rigerszki, Elena Rocchi, Jorge Rollán, Thorsten Saul, Ida Sborgia, Alex Schmidt, Bastian Schubert, Kerstin Schwindt, Roberto Sforza, Alexandra Spiegel, Lidia Tomaro, Nuno Torres, Laura Valentini, Umberto Viotto, Henrike Wettner, Waldemar Wilwer, Sabine Zaharanski, Gabriele Rotelli, Guile Amadeu, Nuno Almeida, Felipe Bernal, Milena Boxberger, Daniel Burston, Giovanni Cardone, Carolina Civarolo, Carla Cruz, Daniel Erfeld, Elizabeth Farkas, Ina Fertig, Rafael Galvis, Sergio Leone, Cassía de Godoy Lima, Abelardo Gómez, Enno Hergenhan, Valentin Kokudev, Michael Kührt, José Manuel LaTorre, Javier Logreira, Desirée Mann, Dirk Mayer, Sylman Mirza, Christian Molina, Kate Moore, Maria Pierres, Ligita Nicgale, Lina Parra, Catalina Pinzón, Jordi Roldán, Nuno Rodrigues, Miguel Romero, Miguel Sánchez, Jana Scheifele, Anna Stoppani, Rocco Tenca, Martina Viganò / **CLIENT:** Hafencity Hamburg GmbH / **LANDSCAPING:** WES & Partner Landschaftsarchitekten, Wehberg, Schatz, Betz, Kaschke, Hamburg / **ENGINEERING:** Körting Ingenieure, WTM Ingenieure / **BUILDER:** FA. Boy Mann

CERAMIC: Solid terracotta tiles of different colours and sizes 21.5 x 21.5 x 10.5 cm; the same elements on walls protruding in different ways

Gonzalo Silva Moreno
Javier Mira Peidró
Institute of Ceramic Technology (ITC)

Innovative products in the ceramics sector

The products that are explained in this section were chosen from among a wide range of innovations being developed in the ceramics sector in recent years. The selection criteria was based on the suitability of these products for use in architecture, either for their novelty or because they comply with the requirements of the new Spanish Technical Building Code.

Translucent ceramic [1]

These are translucent vitreous tiles obtained by means of sintering frit particles and come in a range of 76 different shades. This rectified tile comes in the following sizes – 29 x 29 cm, 26 x 56 cm and 44 x 44 cm with thicknesses of 7, 9 or 12 mm. They can be hung on walls by means of metal or wooden frames onto which the translucent tile is fixed using structural silicone glazing, fastening systems (hook and loop or similar) or mechanical systems. For technical flooring, this translucent ceramic requires reinforcing with polyvinyl butyral (PVB) glass lamination using any of the existing systems in the market for this kind of flooring.
An interesting product[2] based on this translucent ceramic is the fibrocement block with translucent tiles on each of its two faces. Available sizes are 41 x 41 x 8 cm and 20.5 x 41 x 8 cm. Both formats allow the possibility of interior LED lighting.

Photovoltaic panels

The current awareness of environmental issues like alternative energy sources is reflected in the ceramics sector by the two systems put forward to allow photovoltaic cells to be integrated into tiled façades.
The first of these[3] has the generator circuit encapsulated between the tile and its protective laminate. The protection is made up of a triple layer of TEDLAR® (polyvinyl fluoride) / PET (polyethylene terephthalate) / TEDLAR® with thicknesses of 0.025, 0.075 and 0.025 mm, respectively. While the outside TEDLAR® film offers good resistance to the elements and the inside layer adheres well to the encapsulant, the middle layer provides electrical insulation and mechanical stability. Laminate of this kind also is less permeable to atmospheric moisture.

1. ESMALGLASS.
2. P-BLOCK ROCA.
3. FRITTA and ISOFOTÓN.

Ethylene-vinyl acetate (EVA)) is used as encapsulation material to impregnate the circuit made up of solar cells. Its most important features are its low water absorption and its transparency. The top layer protects it from ultraviolet light.
The image shows the installation produced with the technique of integrating photovoltaic cells with ceramic tiles in Freising (Germany). The project[4] consists of a 1-km long noise barrier on the motorway connecting Munich with its airport.
The modules that have been installed are made of porcelain stoneware[5] tiles with photovoltaic cells integrated with the previously described technique. A total of 6,750 ceramic modules measuring 100 x 50 cm were put in place to generate 338 kWp. The modules provide highly efficient acoustic insulation as they combine their assembly advantages with costs that are lower than the typical tempered glass panels.
The second system[6] consists of anodised aluminium frame panels integrating square photovoltaic cells made of crystalline silicon. These panels are adapted to coexist on the façade with porcelain stoneware tiles. They come with upper and lower horizontal profiles adapted for hanging on the outer skin of the ventilated façade as easily as the stoneware tiles. The high-efficiency cells are totally encapsulated in EVA and protected against dirt, moisture and knocks by a special pyramidal anti-reflective tempered glass facing and a black TEDLAR® backing for better blending and to ensure total watertightness. The facing of each photovoltaic panel is made of special pyramidal glass at 45° and with low iron content. This allows it to perform better in a vertical position. In practice, it has been shown that pyramidal glass gains between 10 and 15% more incident solar radiation, leading to increased energy output fed to the grid.

Thermoacoustic partition walls[7]

This is made up of multi-layered tiles alternating insulation and ceramic panels. They can be used to build partition walls with a high degree of acoustic insulation (up to 60 dB) and fire resistance (up to 120 dB). The tiles come in three models with thicknesses of 6, 10 and 13 cm. Installation is easy given the tongue and groove joints on each piece that allow use of adhesives (typically known as tile cement) for laying or hanging ceramic tiles.

Ergonomic paving[8]

This line comes under the framework of the development of new materials and ceramic systems with ergonomic properties that contribute to improve users' quality of life. It proposes the design of products following Universal Design guidelines and taking biomechanical criteria into account. Its aim is to encourage accessibility to persons with reduced mobility (PRMs) and the elderly and increasing user comfort and safety while promoting the extension of the use of tiling to new areas such as pavements, walkways, squares or urban furniture. The elderly are considered part of the population group at most risk from suffering falls. Factors behind this problem include visual impairment, diminishing cognitive faculties, lessening of strength in legs and loss of balance.
Paved surfaces should not enhance any of the deficiencies that come with age, which is why these pavings have properties that ensure continued grip (in addition to ensuring adequate grip for walking with crutches and sticks) and changes in colour that warn of danger and inform passers-by of raised areas not greater than the toe clearance of elderly people. Once again, by incorporating design elements that convey the

4. The project was financed by the European Commission DG TREN
5. PAMESA.
6. TAU CERÁMICA.
7. TERRA-CUITA.
8. TAU CERÁMICA. CERGOCIVIS Project: co-financed by IMPIVA, PROFIT and CDTI. (*) Cergovis (Ceramic Systems with Integrated Ergonomic Functionality and Universal Design Installable on Different Surfaces) is a project with technological advice from ITC - Institute of Ceramic Technology and IBV - Biomechanics Institute of Valencia.

idea of safety to older people, a new concept is created: emotional accessibility.
Toe clearance is a design parameter for paving obtained from clinical studies on population. It fundamentally defines the minimum separation between the shod foot and the horizontal surface produced during the shift in support while walking. A paving with ideal toe clearance has a maximum raised surface and a design appropriate for not having to modify the normal gait or, in the worst possible case, to not cause tripping with the well-known consequences such as broken hips, wrists or serious bruising.
This project has developed criteria for designing ceramic paving tiles providing adequate toe clearance, based on integral covering. Therefore, biomechanical and ergonomic criteria join forces with the properties of ceramic for the first time to offer a product with improved functions.

Outdoor paving tiles [9]

This family of products covers long-wearing paving tiles for outdoor areas that convey safety and comfort to the citizen.
The possibilities of these products respond to a wide range of urban projects and are therefore not bound by the rigidity of a single, regular format.
When we walk, we all try to develop a gait that is comfortable and safe depending on the speed we wish to keep. For this reason, this line of products incorporates design parameters that convey safety and durability, and in so doing, add emotional functions. It also complies with the recommended toe clearance parameters.
Strength is given to the tile by modifying the geometry of the back (rib) based on the calculations of stress by means of finite elements analysis. The aim is to reinforce the most vulnerable areas asymmetrically, even under the most unfavourable laying conditions.
These geometrically-modified paving tiles can be laid in heavily transited urban areas such as pavements and walkways given that their design and resistance makes them particularly suitable for heavy pedestrian use or even with occasional motor vehicle transit. They can be adapted to a wide range of products and are available for paved areas in a great variety of patterns, colours and textures.

Non-slip paving solutions [10]

The Spanish Technical Building Code Section SU-1 Safety where there is risk of falling, Paragraph 1 regarding floor/paved surface slipperiness includes requirements related to floor/paved surface slipperiness in buildings and areas around hospitals, schools, businesses and public places.
The phenomenon of grip can be explained by two basic concepts: adherence on contact, associated with the interaction of the roughness or micro-texture of the paved surface with footwear and adherence by penetration/deformation, associated with the interaction with raised areas of the paved surface and the absorption of energy through deformation of the sole. Depending on the surface characteristics of the paving and the presence of intermediary fluid, both mechanisms can coexist. The former is more typical of dry surface conditions while the latter occurs in the presence of fluids that create continuous films on the surface.
In order to modify this behaviour, action can be taken either on the surface texture or in the design of textured patterns. Increasing surface roughness tends to im-

9. TAU CERÁMICA. CERGOCIVIS Project: cofinanced by IMPIVA, PROFIT and CDTI. (*) Cergovis (Ceramic Systems with Integrated Ergonomic Functionality and Universal Design Installable on Different Surfaces) is a project with technological advice from ITC - Institute of Ceramic Technology and IBV - Biomechanics Institute of Valencia.
10. VERNIS. GrypSystem.

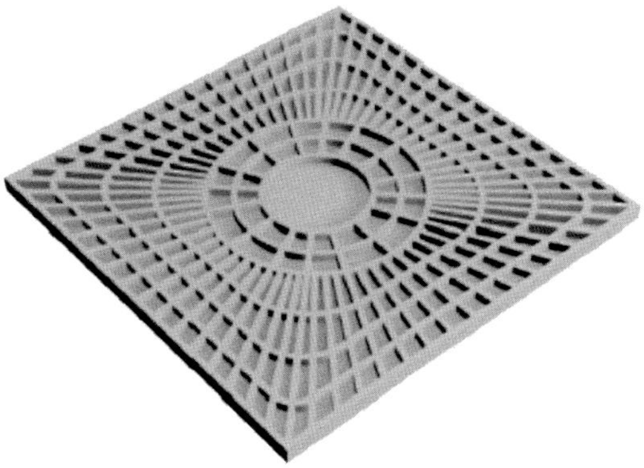

prove grip (mainly in dry conditions) although it can also increase the tendency to retain dirt. On the other hand, although the use of raised non-slip surface textures allows grip to be increased without altering the hygienic properties of the paving, their presence alters gait and inconveniences the user. Consequently, this solution tends to only be used where safety takes priority over all other conditioning factors (industrial use, pavements, etc.) In order to resolve this issue, an application has been developed that allows the surface finish of any paving tile to be modified without altering its appearance, and therefore raising the non-slip performance to meet the requirements established for any kind of premises or use.
Its non-slip effect is based on the creation of a surface micro-texture that improves adherence through both contact and penetration. This may be regulated according to the established needs by modifying the vitreous varnish. Given that the large crystal size and refraction index is similar to those in the vitreous stage, the surface application is practically transparent and does not modify the aesthetics of the product. It can be used during the different routine processes (Red stoneware, glazed porcelain stoneware, technical porcelain stoneware, etc.) prior to firing.

Radiant ceramic [11]

The functionality of the system resides in the capacity of the flooring to transform electrical energy into heat. In order to achieve this, a system has been designed comprising the following elements: functional tiles, conductor joints, thermostat, transformer and differential switch. The following is a description of the system and its main elements.

SYSTEM: The system is basically made up of a set of functional tiles, joined with special grout, that when connected to a low voltage power source heat the ceramic surface. In the figure, the installation is given in which it can be seen that all of the tiles are fed. Although the system works with alternating current, we have represented the wires with different colours to show that each tile must be connected to the two stages of AC.

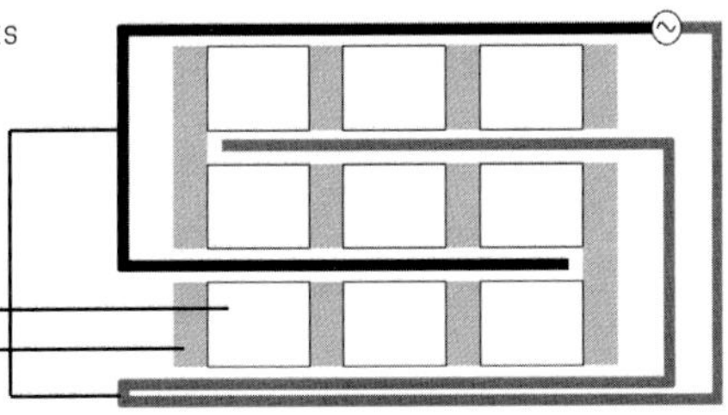

TILE: The tile design has the purpose of allowing heat to be generated by electrical energy. The frit and final glaze composition has been optimised so that resistance values of up to 103 ^{c} were achieved on the surface of the glazed ceramic tiles.
The developed ceramic glaze gives off heat once a low power differential is applied to create the desired feeling of warmth. It has been observed that the temperature of the tile depends on the power applied and that a higher voltage gives a greater increase in temperature.
It should be pointed out that the total installation was done to maintain the safety conditions that allow risk-free use. It is for this that the ceramic tile with conductive glaze is covered with a layer of normal glaze that is naturally insulating.
This insulates the top of the tile whereas the lower part is insulated by the base, which also serves as a storage heater while being an insulator.

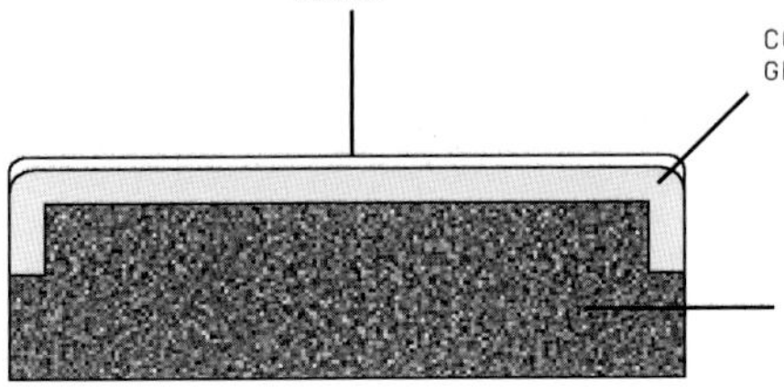

GROUT: To allow the current to reach all of the tiles, a conductive polymer grout is used. It is then covered with an insulating grout to maintain installation safety at all times. Figure 20 shows the complete bonding of two tiles.

11. ROCERSA: Project carried out in collaboration with Roig-Ceramica S.A. (ROCERSA), Tierra Atomizada, Colorrubia España S.A. and Jaume I University.

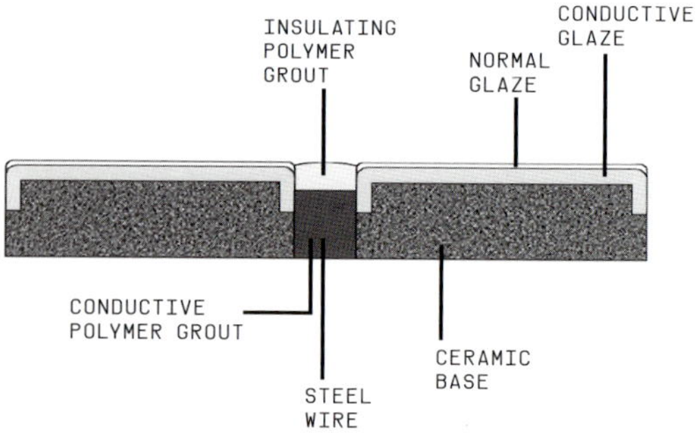

DOMOTICS: Some developments are proposed in this section where tiles, in addition to their natural ceramic properties of resistance, hygiene, the possibility of adapting the surface, etc, have new functionalities that can be complemented with current home automation systems. We highlight two projects that expand this functionality.

The first uses highly conductive materials. If correctly designed, these allow circuits to be closed and a signal to be sent which, once processed, can work the way we want. The application of these materials is by means of PDP (Physical Vapour Deposition) permitting very fine, highly conductive layers to be placed over the tile surface. Use of masks enables the desired circuits to be made with the thickness required for conductivity purposes. These capacitors can be placed on the face of the tile or on the back.

The second project is a development of tiles with sensors on the back that give control of different devices by pressing on the tile. Both their positioning on the wall at a suitable height and large icon with Braille texts contribute to integral access following Universal Design principles.

12. FRITTA and TEKNIKER.
13. TAU CERÁMICA. CONFORTILE Project cofinanced by IMPIVA, PROFIT and CDTI. CONFORTILE is a project with technological advice from ITC - Institute of Ceramic Technology and IBV - Biomechanics Institute of Valencia.

Mila Payá
Institute of Ceramic Technology (ITC)

Experimental Zone: Aesthetic currents and ceramics

Ideas have been emerging from different design forums to respond to the current situation we live in. They are concepts that range from the manufacturing of everyday products in a globalised setting to intimate and unique domestic universes and include trends such as humanitarian action, sustainable development, ecological design, product and service design in addition to subjects related to senses, well-being and personal expansion.
These concepts have motivated a response in line with the times through the design of ceramic products with the latest technology that allow ceramics on offer to be widened and its uses improved both for the public and private spheres.
Alicer, the department of design and architecture at ITC analyses current needs and studies responses through the design of ceramic products. The work involved has several stages beginning with data obtained on the present situation. A group made up of both in-house and external designers work to find an appropriate solution for each problem, need or expectation of the end user.

CON LOS PIES EN LA TIERRA
(FEET ON THE GROUND)
MILA PAYÁ. ALICER

Naturalness

Nature is still an endless, fruitful and stimulating resource for both artists and designers. We continue to seek natural connotations in our immediate surroundings. Whereas recently this nature was transformed to become functional - rustic and functional - as recently we would have defined this line of work, it is no longer enough. We need to come even closer to it. This approach to nature is no longer through interpretation or representation; we get really close until we feel immersed in it and we live with it. To do this and to capture its essence, we must study its structures, growth, rhythms, colours, etc.
This becomes a truly fruitful exercise. Architects and designers coincide in the use of vegetation in its most natural and live state, considering it for cladding, moss walls or turf lamps that change with the passing of time, like nature itself. The ALICER creation, *Con los pies en la tierra* (Feet on the ground) aims to assimilate vegetation into a terracotta paving of circular tiles made by hydraulic cutting. The project is conceived as an alternative for both interiors and exteriors of dwellings.

KU
SIMONETTA CARTA, JUAN PAYÁ

VELADURA (GLAZE)
JAVIER MIRA

DESCORAZONADOS (DISHEARTENED)
PILAR ARIÑO

LEVEDAD (LEVITY)
MILA PAYÁ

Elegance and lightness

Elegance: the absence of the superfluous; the essential; playing with light and shadow; aerial structures reaching out to nature and living with it through elements; seeing and feeling water, air, bright light and reflections.... It is an exercise in synthesis - a challenge for an artist who must capture timeless and unquestionable beauty. It is the triumph of line, subtleness, transparency, style and poetry.
To create this line, the designer Simonetta Carta collaborated with glass designer Juan Payá to bring forth the KU project in ceramic and glass. "Zen comes from the centre, from the vacuity, to penetrate the nothingness. KU is a symbol that combines the visible with the hidden, the simple with the profound, the full with the empty, the nothingness that contains the essence of everything."
On the other hand, *Veladura* (Glaze) represents the immaterial, the inexistent; it gives colour to air. The project composes and creates abstract, aerial and mobile fractal geometries that appear to fall into place randomly and spontaneously to form different, unrepeatable graphic solutions, the common aesthetic feature of which being the enveloping effect they have on anyone who contemplates them.

Passion, excess and voluptuosity

Beyond elegance lies passion. It is a search for emotion through references from other cultures – the exotic, luxury, opulence and splendour. Value is once again given to craftsmanship, relief decoration, crochet and lace, silver and goldsmithing, precious materials and stones, the triumph of colour and voluptuous shapes. Theatrics and dramatisation turn the mundane into scenes of fantasy. Baroque solutions turn everyday objects into luxury, as is the case of drink cans transformed by metal relief decoration, or the Monsieur chair dressed fancifully with an exuberant train. These concepts have led to the *Descorazonados* (Disheartened) project. This is a decorative covering that draws on imagery of ways "to break a heart", making reference to people's states of mind and emotion.

Magic, fantasy, dreaminess

This line, responding to a spirit of dream and restlessness, recreates surprising, changing effects and products that stimulate the senses to trigger emotions that relax the thoughts; emotions that create pleasant ambiences where dream and reality converge to construct an oasis for the mind and relaxation for the body. These are products made to be contemplated, and ambiences for dreaminess. It is about feeding body and soul.
In the Pompidou Centre exhibition, Le Design Aujourd'hui, the section devoted to the senses showed several projects, among them the giant trees with branches of light by Arik Levi, mobiles made with perfume bottles, or the methods used in Ferran Adriá's culinary workshop. In order to illustrate this current, a ceramic material that is uncommon as floor or wall tiles was used – a translucent porcelain with natural shapes on the front and a system of LEDs on the back to produce a motif recreating fireworks as backlighting for the tile. This system is proposed as cladding for double-skinned ventilated façades that can change their aspect from day to night.
A second proposal within the line "Magic, fantasy and dreaminess" is a cladding where dry glaze is applied to create the effect of raindrops; an effect that varies depending on the angle of incidence of sunlight leading to changes in light and shade as the day goes on. *Zènit* (Zenith) is a project based on the idea of using a flaw in ceramic that becomes an effect if its development is controlled. In this case, the exercise not only aims to create a unique effect but to make use of an aesthetic expression that is purely ceramic.

ZÈNIT (ZENITH)
AROA GARCÍA, NOELIA COLL. ALICER

ANÉMONA (ANEMONE)
CRISTINA SERRANO. ALICER

LÁGRIMAS DE COCODRILO (CROCODILE TEARS)
SILVIA RODRÍGUEZ. ALICER

Examine, question and explore

The aim is to explore the possibilities offered by new materials to obtain new forms, applications and performance by considering alternatives to already existing ones. Examining limitations and surpassing them is an exercise in risk and requires courage. Many professionals feel comfortable with this, creating unique, daring and unrepeatable products such as a hotel under the sea, a building in the shape of a mountain of roses or a stadium inside a turf volcano; these are projects that have been finished or are in progress and prove not to be utopias but reality. The *Anémona* (Anemone) project is a large flower and involves a technique to bend ceramic tiles known as thermoforming. It is an urban furniture design where the petals are seats joined together at the centre by means of a concrete bollard with a ceramic cover cut out by water at high pressure. Ceramic as a material for wall coverings is associated with the idea that a product must be lasting and unchanged. However, this virtue can often become monotonous. The *Lágrimas de cocodrilo* (Crocodile tears) project puts forward a system of tear-shaped, complementary tiles that can be used as a decorative element to revitalise already existing wall tiles or as a solution for creating superficial structures to the user's taste. As a third option, still under the same line of proposals, is the project known as Vidas paralelas (Parallel lives). This is a dual-function tile that can be both wall covering or trays or plates. A shallow recess marks the space where food and cutlery are to be placed. It is an attempt to examine the industrial process involved in making floor and wall tiles in order to create products that offer more possibilities for use, without altering the process.
The tiles come with a hanging system that allows them to be used and then easily returned to their place on the wall, thus turning the wall into a mural that changes as different tiles are being used.
This section ends with the project called *Punt* (Stitch), which proposes a system of rectangular, drilled tiles allowing them to be joined to others, as well as a cross-stitch pattern to be traced on them, to create a curtain or room divider.

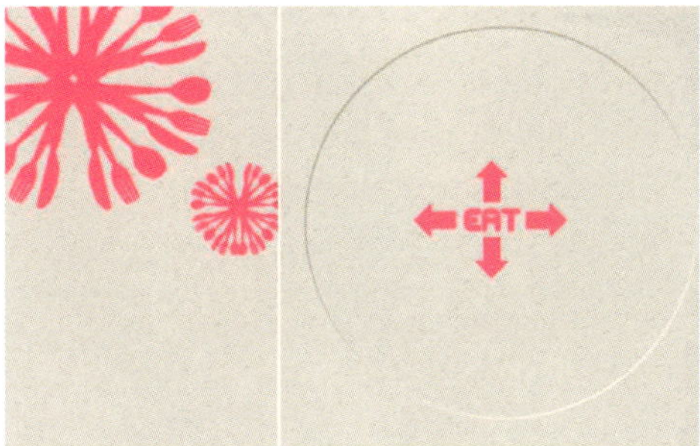

VIDAS PARALELAS (PARALLEL LIVES)
AMPARO BELTRÁN

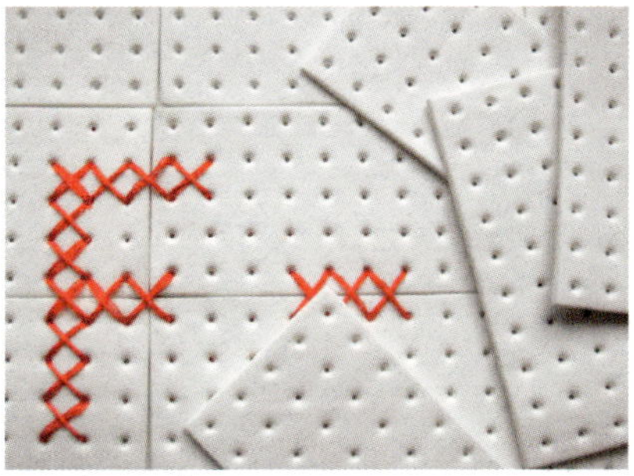

PUNT (STITCH)
ÓSCAR VALLS. ALICER

Illustrate, narrate and represent

Graphic illustration as ornamentation is a resource for creating different looks in one product line. Walls, floors, furniture and even the human body can serve as a support to illustrate or recreate landscapes, floral motifs, ornaments from different artistic periods or characters from fairy tales and legends that, in turn, become the leading threads of an ever more popular aesthetic current.

This project has the aim of giving different perspectives of ceramics through the work of the graphic designers participating in this section. They create illustrations on floor tiles as if they were paper using new systems for digital printing on ceramic products. In this way, the aesthetics of comics and graffiti, or manual artistic techniques such as watercolour, tempera, pencil, marking pens or collage are transferred onto the tile surface.

An example of the popularity enjoyed by this trend is the Hotel Fox in Copenhagen, which opened with the aim of appealing to a younger clientele. For this purpose, the décor of the rooms was given to different artists. These young designers decorated and painted each room freely, without restrictions.

Another example of this line is applied by Adidas. They give buyers of their trainers a paint set so that they can be decorated to the user's taste.

Two illustrators who have carried out these kinds of projects have collaborated with ALICER in this line. The designer Rosina Reig created *Salamarcanda* (Samarkand), illustrating the novel "The One Thousand and One Nights". It has six illustrations that can be used as floor or wall coverings.

Adrián Blanca Méndez produced a design inspired by urban activity and illustrating contemporary society.

ALICER also contributed with the project titled *Gráfica en construcción* (Graphics under construction), a system of tiles that can be moved inside a panel to create different graphics.

The project *De lo plural y lo singular* (Of plural and singular) picks up on graphic motifs made by designers in their notebooks either consciously or as musings. These motifs were collected from the notebooks of designers from the Department of Design and Architecture at ALICER ITC and were laid out at random giving the effect of a wall that is designed with day-to-day spontaneity.

SALAMARCANDA (SAMARKAND)
ROSINA REIG

ARBEITZEIT (WORK TIME)
ADRIÁN BLANCA MÉNDEZ

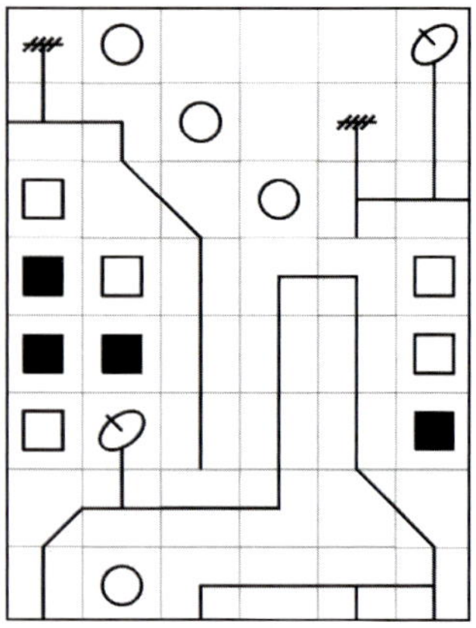

GRÁFICA EN CONSTRUCCIÓN (GRAPHICS UNDER CONSTRUCTION)
CRISTINA SERRANO. ALICER

DE LO PLURAL Y LO SINGULAR (OF PLURAL AND SINGULAR)
ITC DEPARTMENT OF ARCHITECTURE AND DESIGN
ASSEMBLY: ALBA PIÑOT AND LUCÍA ORTIZ

Felipe Pich-Aguilera
Teresa Batlle

Clay Environment Industry Architecture

MESOPOTAMIA. CLAY TABLET WITH CUNEIFORM INSCRIPTIONS. C. 3000 BC

Clay is a raw material with properties and conditions that have led to it being associated with humans since the beginnings of time in their drive to improve the conditions of their lives and the comfort of their homes.
In fact, archaeology acknowledges clay to be a systematic record of the passing of ancient cultures and of their progress over time since clay itself was always a vehicle for technological development in the hands of man for the evaluation of civilisation and expression of culture.

We have recovered small votive figures found in Europe and dating from 23000 BC and fragments of the first pottery containers from 13000 BC found in Japan. At that time they would have been a technological milestone and a radical improvement to the conditions of life for those communities.
Mesopotamia was where a decisive force in the evolution of civilisation took place in around 7000 BC In a fertile valley environment, clay became the prevailing resource for progress. The technology for structural adobe and the process of making bricks that could be piled into buildings were developed that later would blossom into sumptuous murals of vitreous ceramic to represent power and social structure. And above all, it was with clay tablets from 3000 BC that writing developed as uniform incisions in the soft material.

We know that clay has always been linked to peaks of technological development in the evolution of civilisation. In our present time, it is used in the porcelain tiles on the fuselage of space craft, thanks to its great resistance to absorption and extremely high temperatures.
Within this present context however, what is happening in construction? Are we really applying the technical capacity of our industry and the specific qualities of clay to obtain new uses to define our buildings and our way of inhabiting them? Are our typical ceramic processes qualitatively better than those we created in the past?

I really do not believe they are. It seems that there was a fracture in the recent past which led to the whole building sector, being trapped in time for something like three decades.
This fact is a sign that in our contemporary culture, building is not at its core, and though it hurts to say it, we are not at the forefront. This is quite plain to see in the statistical indicators on our sector in terms of resource consumption, rubble and waste production, CO_2 emissions, energy use, but also in terms of expected quality and work-related accidents.
Having reached this point, we should be able to reflect collectively with all the agents involved to try and find ways and shared stimuli to allow us to move forward.
Lately certain symptoms are being felt – and this book is a good example of it – that things are changing. There are surely many reasons for it but, in particular, what can be felt most in this collective evolution is our emerging environmental awareness.

We must know how to take in what is happening around us. One of the genuine traits of our generation is a new appreciation for our natural surroundings; from being considered a source of raw materials to explore, it has come to be appreciated as wealth that we must administer and cherish. This ideological mutation, probably linked to man's rationality in terms of his own survival, is, in our understanding, one of the few globally shared concepts still present after the fall of the grand ideologies of last century. Architecture is beginning to incorporate the basics of this new paradigm, in the knowledge that it still has to make its great contribution. 40% of the CO_2 emissions in the EU are related to building.

Ceramic is a significant element in these terms and its industry seems to be actively identifying itself with the emergence of this new paradigm. This is evident in the fact that, in recent years, a multitude of products and systems, linked directly or indirectly to environmental awareness, have been in development or are more or less operational.

VILAFRANCA DEL PENEDÈS (BARCELONA PROVINCE)
LAW COURTS VENTILATED TILED FAÇADE

We now see the appearance of numerous ventilated façade systems with the presence of ceramic as a new thermal wrapping for buildings to reduce energy demand and to achieve interior comfort. Or the effort to involve thermal inertia and material porosity as in the case of thermoclay in the interests of creating a healthier atmosphere for our dwellings, not to mention the new concepts in solar protection using previously unheard of ceramic elements allowing better use to be made of sunlight while avoiding excessive heating from solar radiation. We could also talk about recent progress on buildings with reinforced ceramic structures – such efficient performance that makes possible amazing thinness and savings on materials. It is worthwhile mentioning that none of these systems try to give a finished appearance; they are evident vehicles of a previous ideology in which their final aspect is the result and not the reason for their use.

We imagine that it is the beginning of a major push by the industry, which will be joined by other industries until the whole building sector is involved.
There is much to be done, it is the effort of a choir made up of soloists; they actually do the same things they have always done but a little more efficiently each time
This is the moment when architects and industry, backed by research centres, learn to work together by setting up real R&D departments for the sector.
We architects know that our profession is not an end unto itself, but an instrument to improve on reality and change it to the limits set by our thoughts. Paradoxically, it is also

the substrate for a truly contemporary architecture, in as far as it expresses and formalises the aspirations and context of the society of its time

We should therefore be aware that industry does not work for us; rather architects and industry in equal parts form a close team working for society in trying to achieve its ambitions and the harmony of its dreams.

We must take into account the fact that transformation and significant changes that are required of us can only come from people - integrated in more complex structures and processes. However, structures at their root tend to continue on with and perpetuate a certain reality rather than transform it.
That is why people and their spirit of innovation and creativity, in a multi-disciplinary dialogue, are the only possible engines for this process.

Therefore, companies involved in innovation processes are not necessarily the largest or appearing to be the most suitable. They are organised around people with enough sensitivity and capability to set up lines of research which, by definition, are uncertain processes and only reach the economic exploitation stage with great difficulty. Indeed, we all know how in other sectors, such as information technology, the largest companies find it hard to innovate and end up buying smaller companies that were able to develop new products.

This is, in any case, an interesting time with great expectations for the industry, techniques, companies, users and definitely for architecture, which must take strength from this new global agreement by giving it a shape that is our expression for the future.

BARCELONA CASA TERESA, DETACHED DWELLING IN COLLSEROLA. FACING FAÇADE WITH TILES

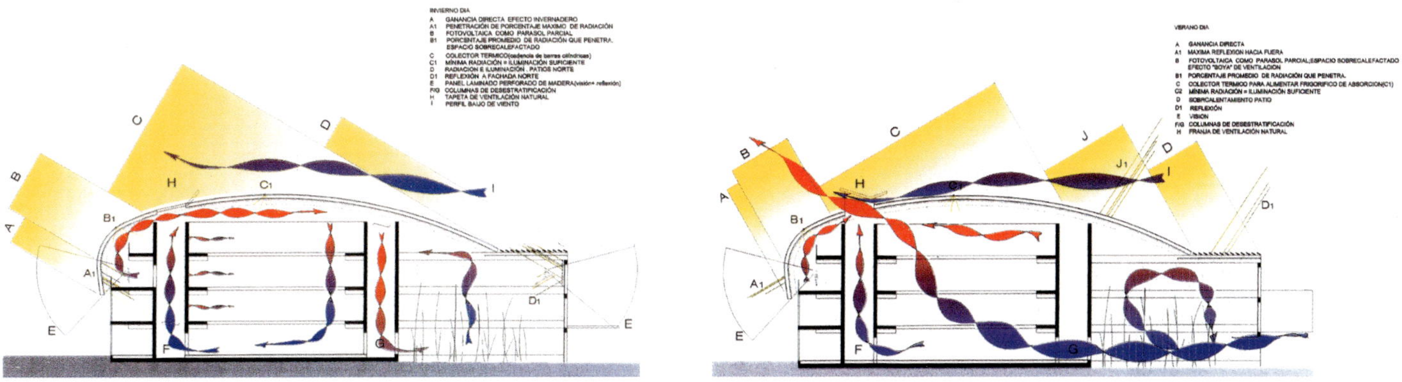

TOLEDO. TELEFÓNICA MÓVILES CALL CENTRE. BIOCLIMATIC PROPERTIES OF THE BUILDING.
LEFT SECTION: WINTER; RIGHT SECTION: SUMMER

Equip XCL
Xavier Claramunt,
Candi Casadevall,
Xavier Medina

Garriga House

Sant Cugat del Vallès, Spain

Garriga house could also easily have been called TUBE House. The house is open in one direction and closed in the other.
The design here was for a house tiled from roof to walls and enclosures but without continuity. The programme is reworked and both walls and roof are covered by a tiled skin. The skin is used to clarify the relation between the design and its location. The rooms are lengthened from one side to the other and the dwelling has a north-south orientation to allow light to flow through the interior from one garden to the other. Split levelling of floors allows the north garden to continue below the dwelling and create a porch.
A transversal section shows that the two parts of the house are united by a void. The façade-roof is faced with a fine surface of white glazed mosaic tiles to reflect the light while the ends are closed with fibrocement, a textured material that gives depth.
Columns and floors are covered by the same continuous tiled façade-roof. One of the sides rests in a gravel-filled well that serves as a rainwater cistern; the other side is suspended to allow light to enter the porch. An angling in the glazed skin heightens the light entering the house and connects it to the adjoining garden outside.

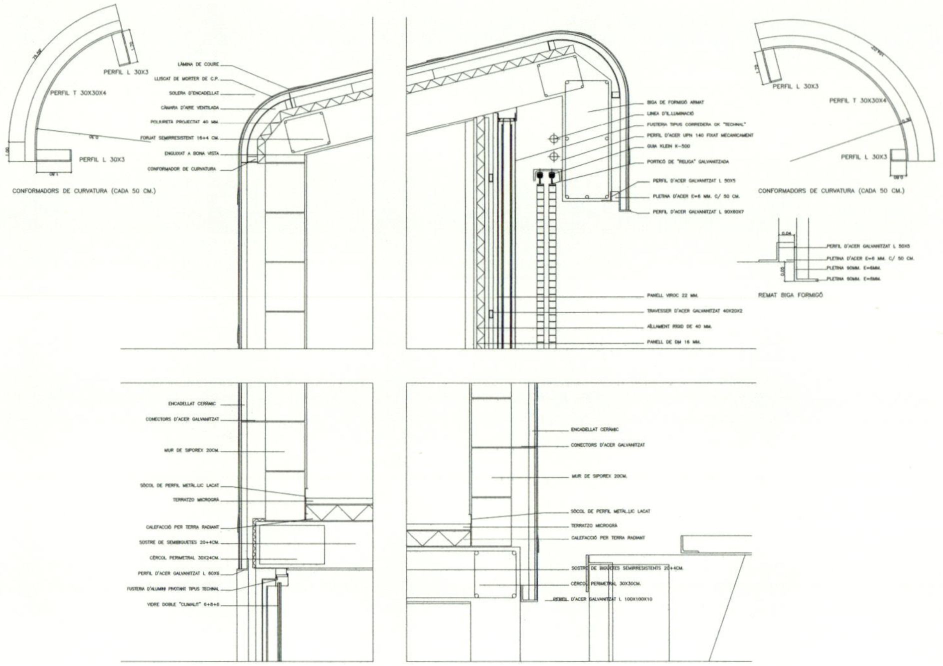

TILED FAÇADE AND ROOF CONSTRUCTION DETAILS

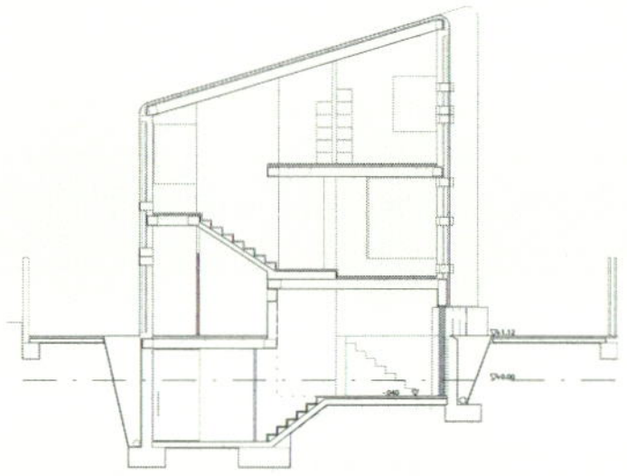

TRANSVERSAL SECTION

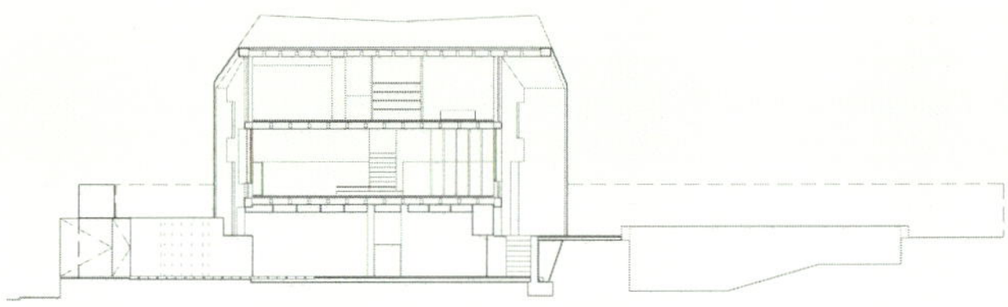

LONGITUDINAL SECTION

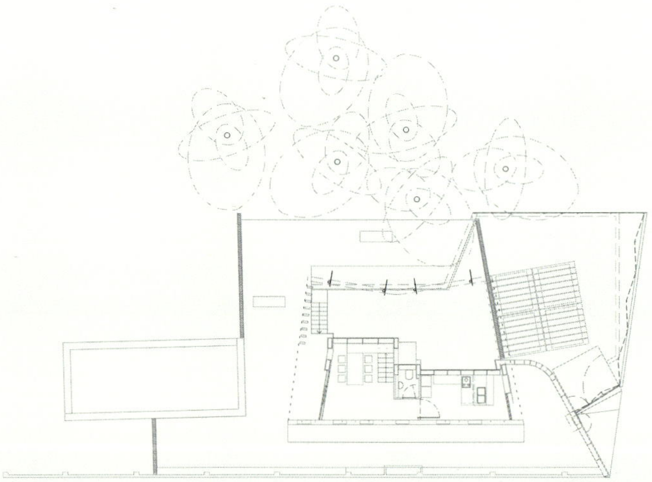

GROUND FLOOR

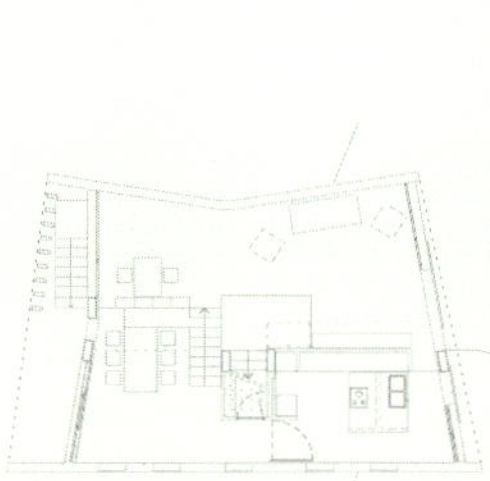

FIRST FLOOR

ARCHITECTS: Equip. XCL, Xavier Claramunt, Candi Casadevall, Xavier Medina / **LOCATION:** Sant Cugat del Vallès, Barcelona / **PROJECT:** 2000 / **CONSTRUCTION:** 2002 / **COLLABORATORS:** Martín Ezquerro, Marc Zaballa, Yago Haro, Miquel de Mas, Joan Cuevas, Oliver Schmidt, Carmen Barberà, Anne-Sophie de Vargas, Pau Rodríguez, Vicky Pons, Ana Azevedo, Adam Mendoza, Alex Ortiz, Hilda Compte, Oriol Bordes, Anna Ramos / **CLIENT:** Familia Garriga / **STRUCTURE:** CROL (Construcción y Rehabilitación Oliván) / **BUILDER:** CROL (construcción y rehabilitación Oliván) / **INSTALLATIONS:** Instalaciones Lairisa / **WOOD WORK:** Carpintería Valdivieso / **METAL WORK:** MAP Cerramientos / **PAINTING:** Pinturas Lluch / **BUILT AREA:** 180 m² / **TOTAL COST:** 210.000 € / **PHOTOGRAPHER:** Giovanni Zanzi

CERAMIC: Ceramic skin facing exterior walls and roof in one white glazed mosaic tiles 5 x 5 cm

Roldán + Berengué
Arquitectes
Miguel Roldán,
Mercè Berengué

Offices and dwellings in the Eixample District

Barcelona, Spain

Tile of Spain Architecture and Interior Design Awards 2006

The space is a horizontal void on the principal floor, crossing a building in the Barcelona's new town district transversally from the street to inside the block.
The building dates from 1864 and was the first on he block.
The space is a perfect rectangle measuring 15 m wide, 35 m long and 3.7 m high with a street frontage facing northeast to the blue light and the back opening onto a courtyard in the centre of the block to the southwest. Here the golden light of midday is refracted into oranges, yellows and blues as it passes through the coloured glass of the enclosed balcony.
Of the five light wells that existed in the original building, there are only four remaining, situated in the middle section of the longitudinal axis of the floor. One of them has a white skylight over one of the entrances. These wells let in light and are windows to the sky.
The floor plan is laid out in three bays measuring 4.5 by 18 m, separated by two 1.5 by 18 m hallways that come together at each end in single rooms that run the width of the floor. The hallway on the right receives yellow light. The one on the left, bluer in colour, makes objects seem pale.

The space can be defined as a container made up of a ceiling plane, a floor plane and vertical planes, some of them structural in nature. Floor and ceiling can be perceived as a whole, creating a series of tiled carpets. The result is a stage constructed with doses of solidity, darkness, depth, height, accumulation, density and detail - the codes of many architectures and of memory.
Several ceilings maintain the original paint from the second renovation the property underwent in 1904. The floors, mosaic in Nolla porcelain stoneware tiles (manufactured by Mosaicos Nolla in Meliana from 1862, colour applied to the tile mass and fired at a high temperature in the style of the Milton company in England) has been restored and some fragments have been reconstructed. Any new intervention to these plans have been kept discreet by using unfinished materials like wood or metal that has been left practically untreated after supply.
Ceramic in the form of porcelain stoneware has been reintroduced in the bar, office and on the terrace and in the form of white bevelled edge tiles on the patio and toilet walls.
The walls are perceived in a fragmented way as a series of parts linked by their direction, but with autonomy of position and form. On vertical surfaces any intervention has been concentrated on perforating and framing to diagonally join rooms that were not connected originally. The result is a layout of rooms which, while maintaining the original structure of the property, are visually linked by new bifurcations like those that one would experience if travelling inside a thick sponge.

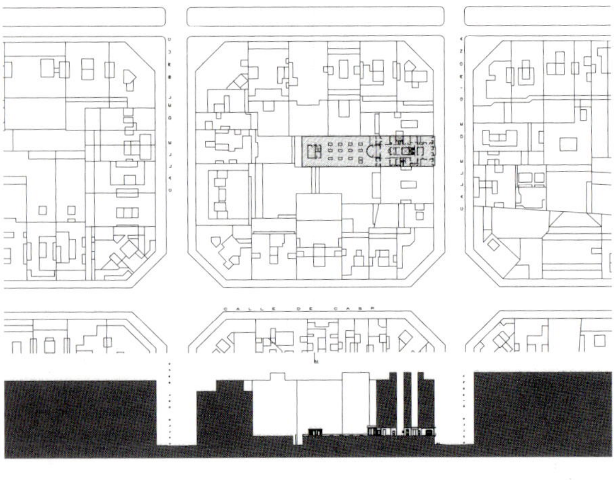

LOCATION

FLOOR. TILED CARPETS. FLOOR TILE TEXTURES

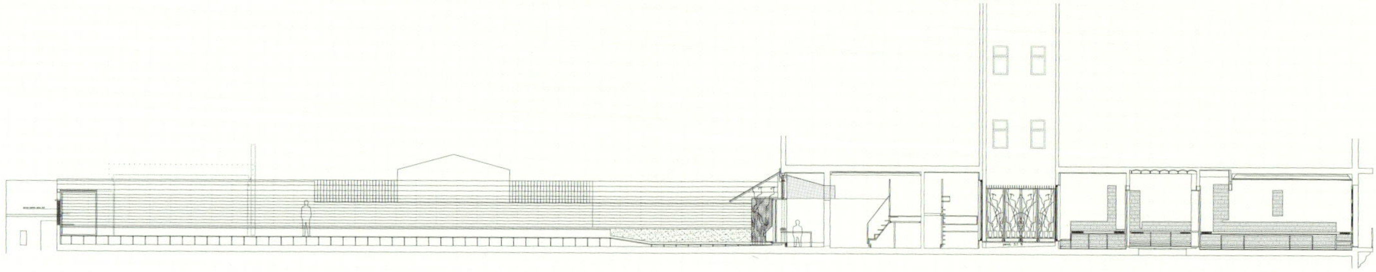

LONGITUDINAL SECTION 1

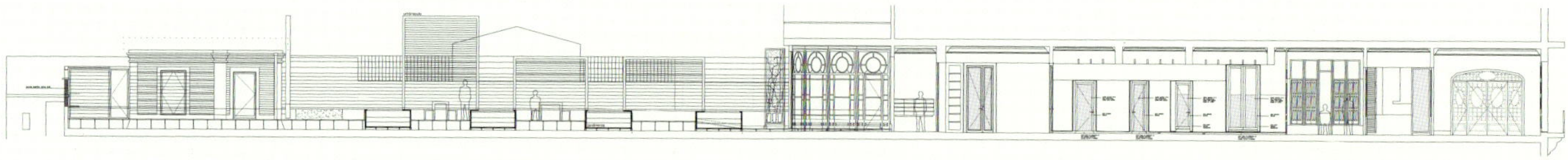

LONGITUDINAL SECTION 2

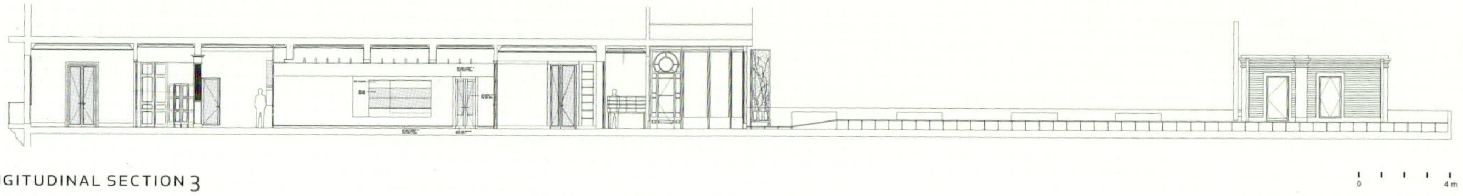

LONGITUDINAL SECTION 3

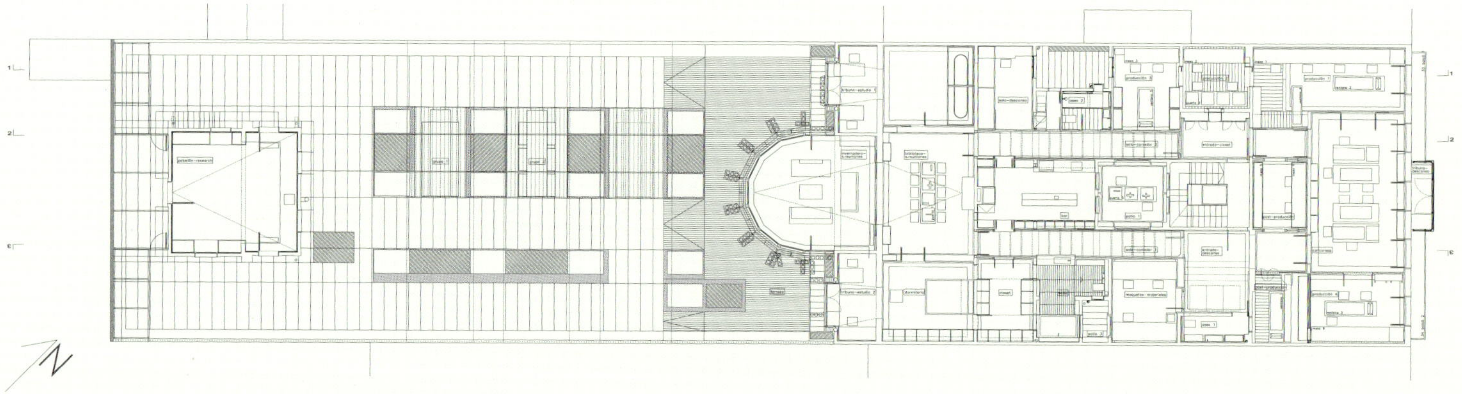

FLOOR PLAN

ARCHITECTS: Roldán + Berengué Arquitectes. Miguel Roldán, Mercè Berengué / **LOCATION:** calle Girona 37, Barcelona, España / **CLIENT:** R+B Arquitectes / **COLLABORATORS:** Sección de Conservación y Restauración del Dept. de Pintura de la Universidad de Barcelona (restoration ceiling paintings); João Camarinha, Miguel Ferreira, Ana Carmo y Jenny Schildecker (arquitectura) / **STRUCTURE:** Eskubi-Turró Arquitectes s.l. / **INSTALLATIONS:** Valdomar Instalaciones, s.l. / **DEVELOPER:** Divisa Barcelona, s.a. / **QUANTITY SURVEYOR:** R+B Arquitectes / **BUILT AREA:** 425m² offices and dwellings + 475m² terrace / **TOTAL COST:** 421.000 € / **PHOTOGRAPHER:** Melissa Bauld and David Morris

CERAMIC: Restoration of Nolla porcelain stoneware mosaic with replacement of missing porcelain stoneware mosaic tiles and framing of each carpet with black sheet steel; porcelain stoneware paving in bar, office and on terrace: APV – Geomusgo Lappate 60 x 30 cm; wall tiling in light wells and toilets: bright white bevelled edge tile 7,5 x 15 cm

Miguel Ayape Bariain, Andrés Martínez Tejada

Artica civic center

Berrioplano, Spain

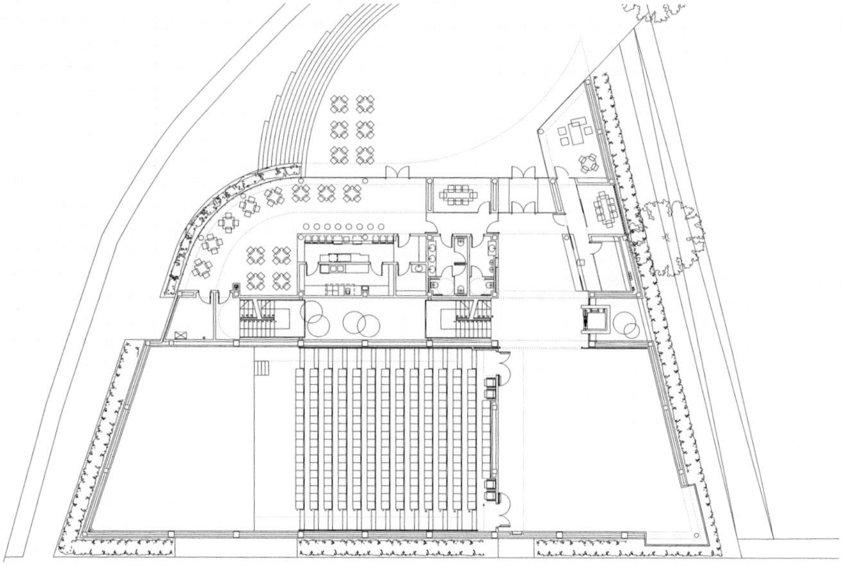

GROUND FLOOR PLAN

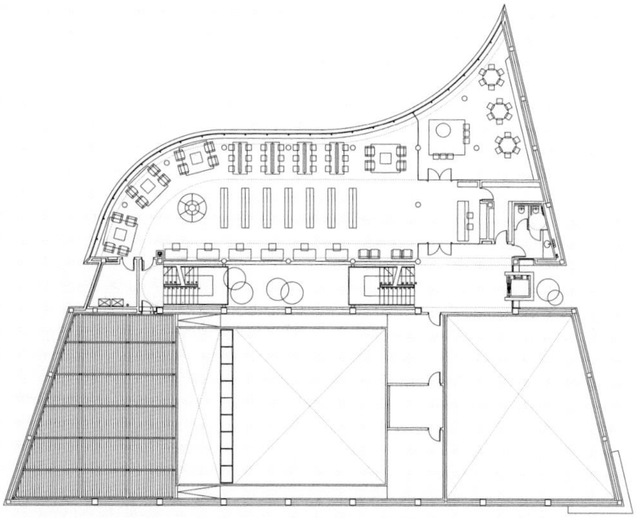

FIRST FLOOR PLAN

This civic centre serves a number of different cultural functions with its theatre seating 248 spectators plus adjacent storage and dressing rooms, library, exhibition room, multi-purpose hall, bar and different service and administration areas.
The centre's location between two public buildings and between two differentiated areas was key to its current use. It stands between the old town of haphazard layout with the traditional architecture of its houses, *caseríos*, and the newer area typified by a geometric street plan set about a central open space.
The building was designed to adapt to the contours of the lot and in response to its functions. The building is arranged in two distinct blocks separated by a central atrium the height of the building like an indoor garden and through which light penetrates into its core. The block nearest to the *frontón* (pelota court), occupying the area of the lot with the most limited views is the theatre. It also takes advantage of the natural slope of the land to allow its public entrance to be on the ground floor, while access to the backstage area and the emergency exit are on the lower ground level but opening onto the street.
The performance area is extended to include the foyer, which is used for other events and exhibitions. At this end of the building, spanning the indoor *frontón*, there are large windows in the corner offering diagonal views over Pamplona and the nearby mountains. These views can also be appreciated from the building's foyer.
The northern face is curved, like a free brushstroke over the urban landscape, and introduces an element that helps to define the small public space in front and to provide a face to the building.

This differentiates the centre from conventional buildings and promotes the public nature of it. The different parts of the complex are brought together through use of materials. Black stone cladding is used on the plinth course, continuing the line created by the paving on the small square in front of the building while the ventilated façade above is clad with large porcelain stoneware tiles. These tiles are held in place by a system of aluminium guides. Inside, the most representative areas and those of double or triple height also feature large ceramic tiles.
Natural light is provided by large panes of glass, in such a way as to let it become part of the building's skin and a defining element of the façades.

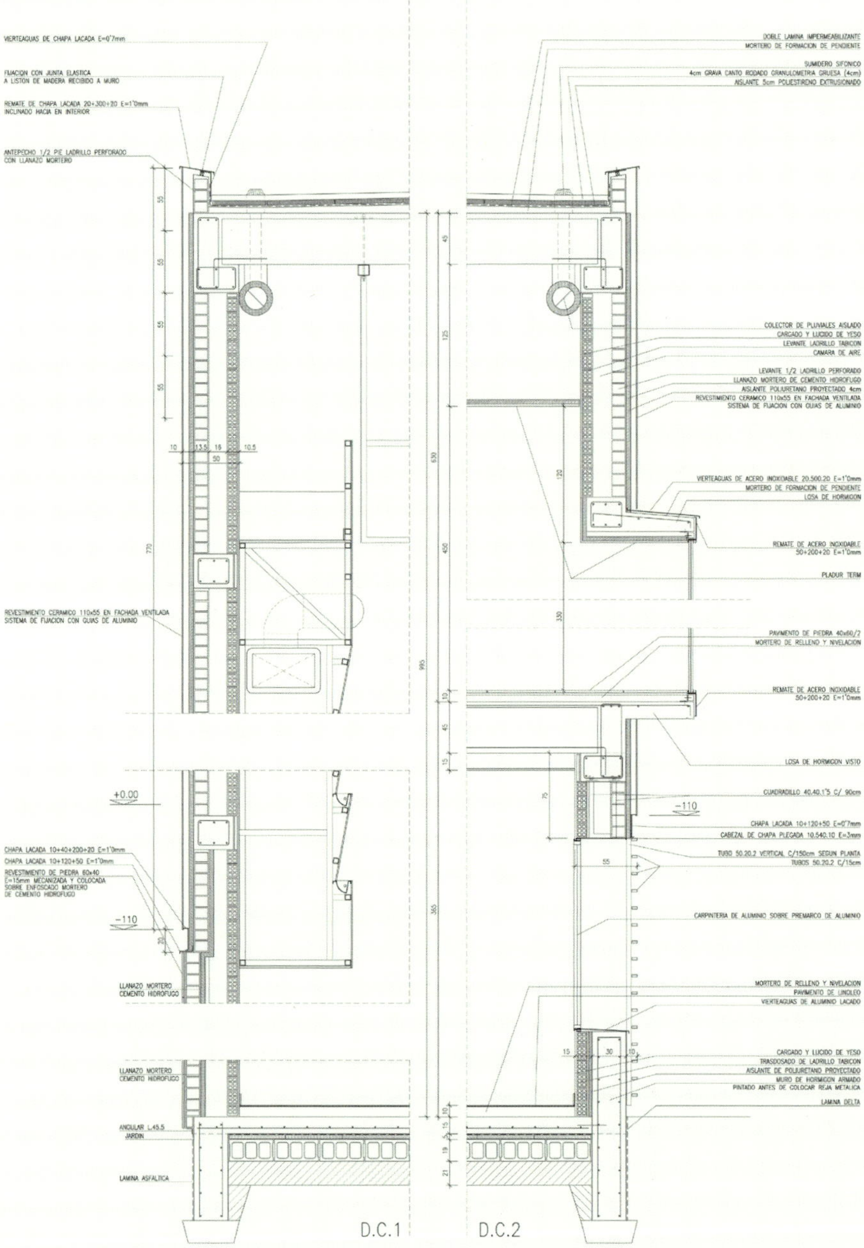

VERTICAL SECTIONS CONSTRUCTION DETAILS

GROUND LEVEL

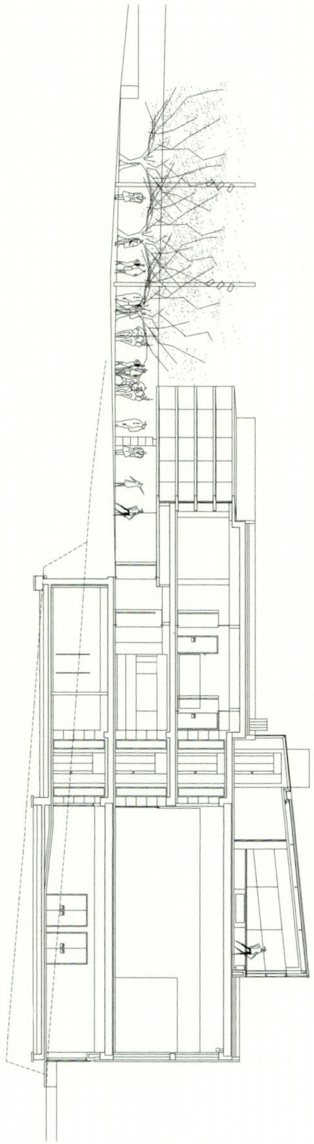

ARCHITECTS: Miguel Ayape Bariain, Andres Martinez Tejada / **LOCATION:** Artica, Berrioplano, Navarra / **COMPETITION:** 2004 / **PROJECT:** 2004 / **CONSTRUCTION:** 2005 – 2006 / **ARCHITECTURAL TECHNOLOGISTS:** José Luis Sola Labari, Stela Fernandez Ayape / **CLIENT:** Berrioplano Municipal Council / **ENGENEERING:** PyP Proyectos y Promoción de Ingeniería / **INSTALLATIONS:** PGI, Instalaciones Arquitectónicas / **BUILDER:** HM / **BUILT AREA:** 2.321 m² / **TOTAL COST:** 1.804.873 €

CERAMIC: Ventilated façade: porcelain stoneware tiles 45 x 90 cm; interiors: ceramic tiles 25 x 36, 20 x 40, 45 x 90 cm

Cloud 9
Enric Ruiz-Geli
Villa Nurbs
Empuriabrava, Spain

Villa Nurbs is an unconventional house in which nature and technology are key elements of its design. Based on the idea of a melting block of ice, the architect has designed a building that withdraws into its interior, housing a swimming pool and winding rooms.
The landscape is conceived as a habitation platform, while the measuring and construction of this landscape is revealed like a 3D topographic image. By optimising building resources by means of CAD/CAM interaction, a wave-like NURBS (Non-Uniform Rational B-Spline) was created based on the climate, geography, external landscape and even the skin of the dwelling.
This skin is reactive and controls energy, privacy, directions and flow of people, and exposure to wind and sun in the different seasons. It is not an abstract design but an overlapping of different diagrams.
The ceramic skin provides a new way of considering the façade in the field of architecture.
It is made up of a series of ceramic tiles designed in such a way that, when seen in their totality, they function like a green wall. Depending on the orientation, this skin has the capacity to close the building to aggressive environmental conditions and to protect it from excess solar radiation, rain or strong winds.

At the same time, it can be permeated by breezes. A net-like system of tensioned cables anchored on the metal structure of the façade serves as the support for the tiles. Both elements define the outer layer of the façade. The painter Frederic Amat worked on some of the tiles by applying glazes. The study of the shape of the tiles was done on a three dimensional model. They were later produced by means of digital manufacturing processes. These processes consist of a relation between the digital design software (computer-aided design – CAD) and the software of the machine that cuts the pieces in wood (computer-aided manufacturing – CAM). The dialogue between the design software and the machine software allowed the virtual model to be produced physically. Ceramic versus the digital age.

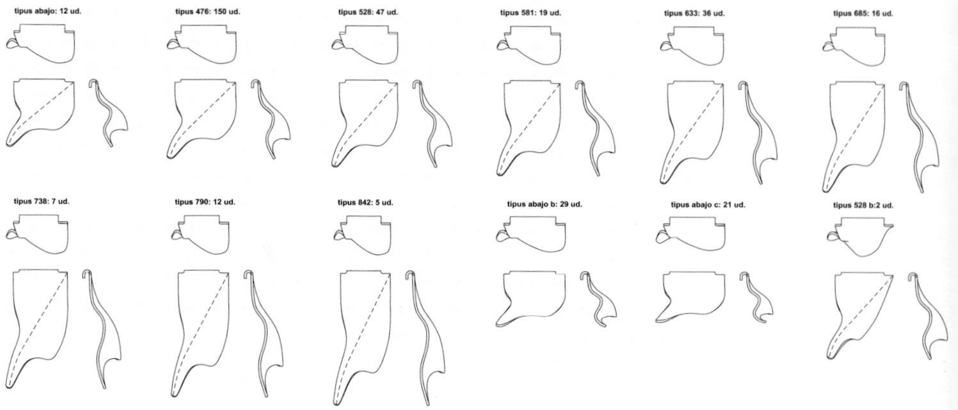

TILES COVERING THE BUILDING

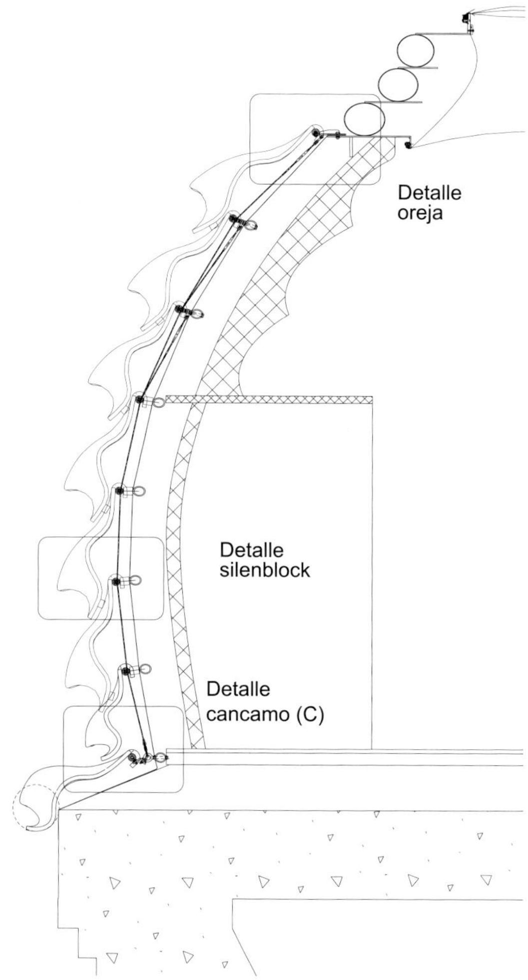

CERAMIC CLADDING TILES

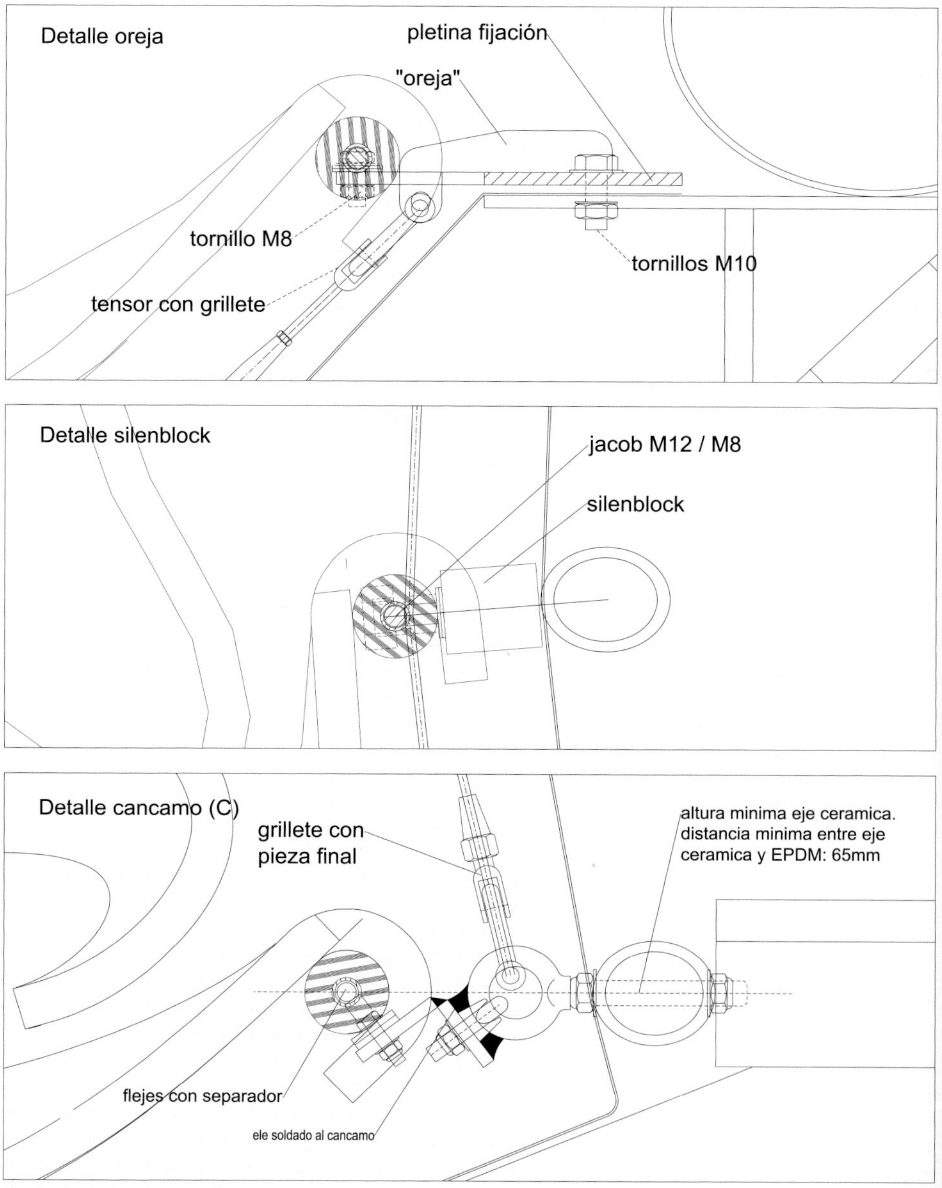

TILE FAÇADE SECTION DETAILS

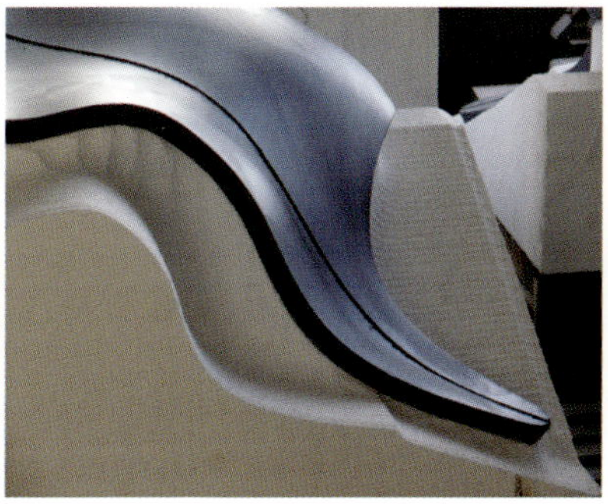

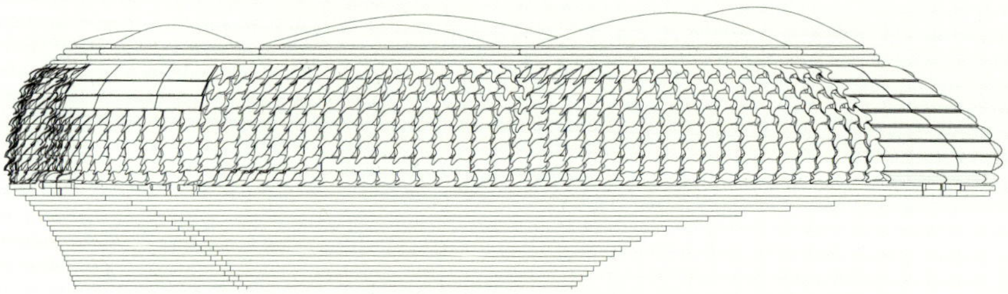

SIDE ELEVATION

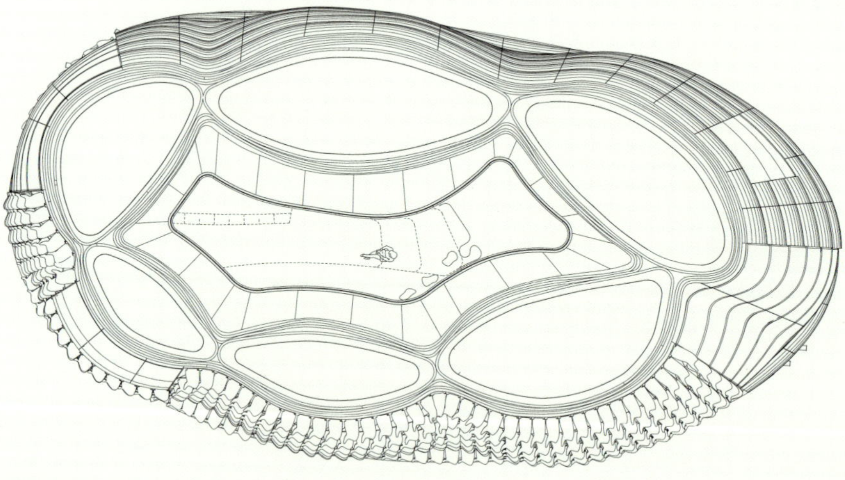

ROOF PLAN

ARCHITECT: Cloud 9. Enric Ruiz Geli / **PROJECT ARCHITECT:** Felix Fassbinder, Jordi Fernández Río / **ARCHITECTURAL: TECHNOLOGISTS:** Daniel Benito Pò, Xavier Badia, Agustí Mallol / **LOCATION:** Empuriabrava, Castelló d'Empúries / **COLLABORATORS:** Miguel Carreiro, Emmanuel Ruffo, Rosa Duque, André Macedo, Ura Carvalho, Hye Young Yu, Marta Yebra, Mae Durant, Angelina Pinto, Randall Holl, William Arbizu, Max Zinnecker, Laia Jutgla, Manel Soler, Megan Kelly-Sweeney, Alessandra Faticanti, Susanne Bodach, André Brosel / **LANDSCAPING:** Jerónimo Hagerman / **FRAME:** BOMA s.l. / **INSTALLATIONS:** PGI Group, Aiterm / **BUILDER:** Obres i Construccions Joan Fustè / **PHOTOGRAPHERS:** Luis Ros, Ramon Prat

CERAMIC: Tiles specially-designed from three-dimensional virtual model and produced by a computer-aided manufacturing process. Assembled over a system of tensioned cables anchored to a metal structure. Glaze applied manually by the painter Frederic Amat

Vicente Sarrablo

Versatile louvres

Tile of Spain Architecture and Interior Design Awards 2004

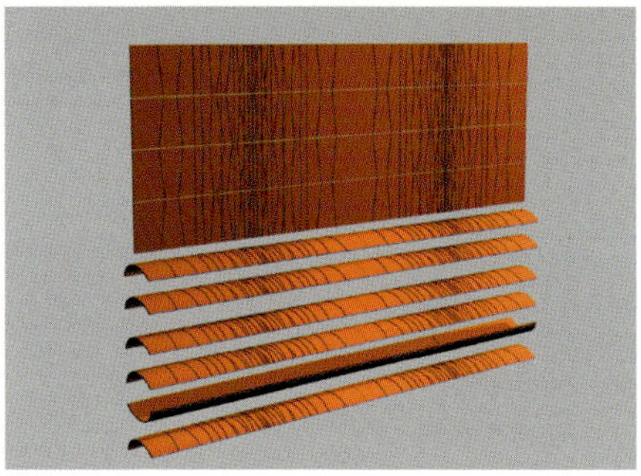

Our Mediterranean climate has always drawn architectural interest to solar control systems by means of latticework and louvres, for the quality of the filtered light they give, the vibrancy added by their presence to façades and for the considerable savings they bring in climate control. Use of a material like porcelain stoneware for these purposes takes advantage of its excellent quality in the face of the elements. They need no maintenance and neither warp nor fade under conditions of humidity or harsh sun.

This product presented a solar control system for façades and roofs based on large porcelain stoneware louvres. The novelty of the design is in its taking advantage of recent tile-bending techniques to obtain a tile with a specific curved profile that allows different uses while providing continuity with the opaque parts of the façade, clad with flat tiles of the same porcelain stoneware. Additionally, curving gives louvres significant rigidity and saves on the normal steel reinforcements needed when installing.

Their main use would be to filter sunlight from glassed areas of façades and intermediate areas between interior and exterior (patios, porches, stands) The same louvres that protect from the sun during the day can provide light at night using its curved profile as lamp shade for bulbs housed in the concave part.

The louvres are so versatile that, when turned 180°, they can be used as a planter box and when several are positioned on a vertical surface, they can become a green wall, an interesting façade with aromatic plants or sedum varieties that need little water.

The rigidity given by the curve of the louvre allows then to be made into resistant components for pergolas and fences. In this way, the advantages of this system go beyond the façade (garden, boundaries).

ARCHITECT: Vicente Sarrablo / **COLLABORATORS:** Jaume Colom, Diego Navarro, Javier Barberá, Wandegar 2001 s.l. / **CRONOLOGY:** casa Barcelona 2006, Construmat

CERAMIC: Ceramic tile, curved fins, 30 cm wide, 12 mm thick, 10 cm tall, 120 cm intervals

Italo Rota

Seafront Promenade

Palermo, Italy

Gold Medal for
Italian Architecture 2006
Public Spaces Category

The project had the objective of returning the seafront to the city by recovering the old promenade with views that it had lost for many years. It was to be a space for recreation, children's games, sport or simply somewhere to lie down and read a book while listening to the sea.
A cycling path was designed to circle the park. built out of green-tinted concrete with pictures in white resin running the length of the route, the path is flanked by a concrete footpath with embedded black lighting fixtures.
Bollards positioned on the road side protect without blocking, prohibiting access to motorbikes and bicycles but giving pedestrian access to the park from any point along the road.
The bollards are made from vitreous terracotta over a concrete structure able to withstand impact.
The design was inspired by the sculptor Laurana's bust of Eleanor of Aragon, kept in the Abatellis Museum. by using the age-old process of reiteration of the figure, new forms were finally created. a number of posts positioned on the main path and the two free-standing sculptures at the entrance are also ceramic with a special gold finish.
The sequence of benches with blue cushions and "lounges" in the rest/contemplation areas, laid out to form "rooms", are also of the same material, while red concrete rugs decorated by young palermitani artists complete the effect.
Additional illumination comes in the form of eight big globes, 160 cm in diameter and mounted on metal posts, that produce a light with a lunar effect.
The sports areas have removable installations on sand or grass and enable all kinds of sporting events and competitions to be held.

GENERAL PLAN

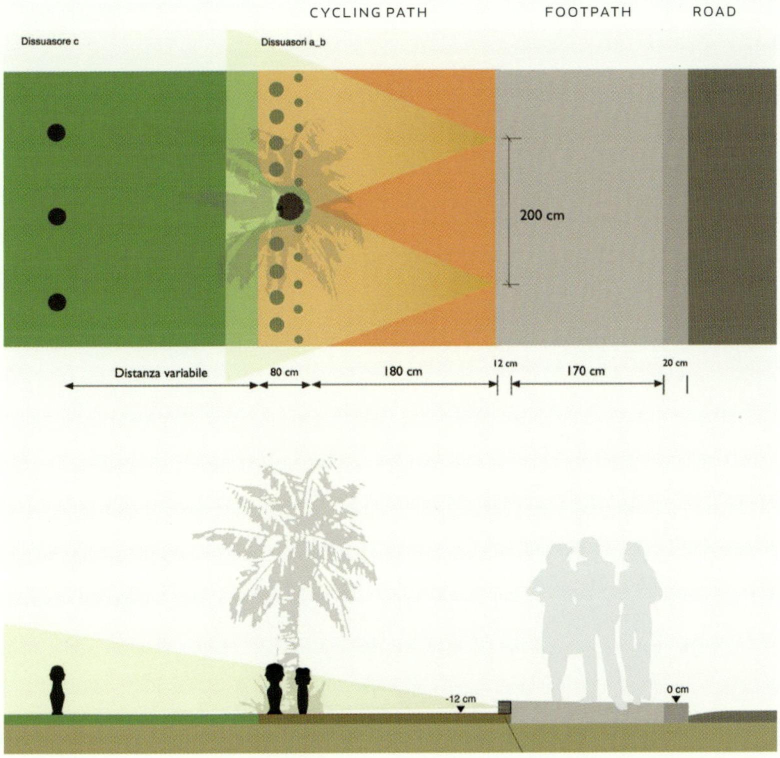

FOOTPATH PLAN AND SECTION

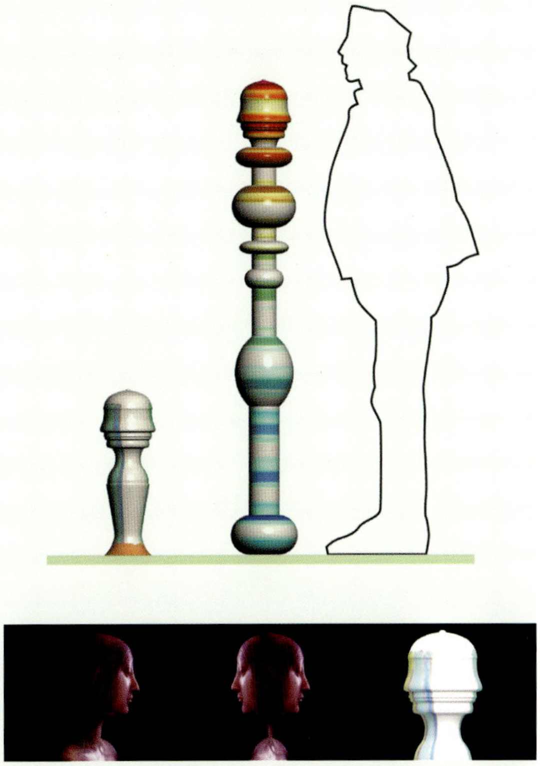

CERAMIC BOLLARDS DESIGN PROCESS

CERAMIC: Unbreakable bollards in vitreous terracotta with concrete structure; benches lined in glazed tiles; central path posts and free-standing sculptures at the entrance with special glod finish

ARCHITECT: Italo Rota / **LOCATION:** Foro Italica, Palermo, Italy / **CHRONOLOGY:** phase I 2005; phase II 2006 / **COLLABORATORS:** Alessandro Pedretti, Francesca Grassi, Rosa Bellanca / **CLIENT:** Comune di Palermo / **LIGHTING:** Alessandro Pedretti / **BUILDER:** Aziende Municipalizzate

Tomás Alía

Bar Discothèque Sol y Sombra

Madrid, Spain

This is a space in the heart of the capital that aims to refine such typical Spanish clichés as are bullfighting and flamenco by representing them in with cutting edge modernity.
The essence of bullfighting has been combined with an innovative and contemporary style, reinventing the look – the ornamental combs, leatherwork or the art of bullfighting on horseback with modern elements. To achieve this effect, energy-saving LED lighting technology allowing colour changing was used.
The furniture looks like it was rescued between wars but it is given a contemporary touch and blends in with the newer elements. Traditional Savonarola chairs are mixed with sofas featuring the purest and cleanest lines.
The most eye-catching element is a wall of digitally-printed tiles running along the length of most of the premises.
The wall shows bullfighters' backs; they are wearing the typical gold embroidered jacket, a reminder of local craftsmanship. The effect is finished off with a sculptural work comprising cubes with colour change LED lights.
The wall behind the bar is lined with leather.
The counter follows the design of those found in traditional Madrid taverns with a zinc countertop. Galactic bottle racks come out of the leather with colour change LED lighting.
Light fittings seem inspired by the Industrial Revolution. A long space invites the visitor to enter, sit down, listen to music or have a drink.
On the left, a giant ornamental comb envelopes and embraces the cocktail bar; its size enhances its feminine splendour.

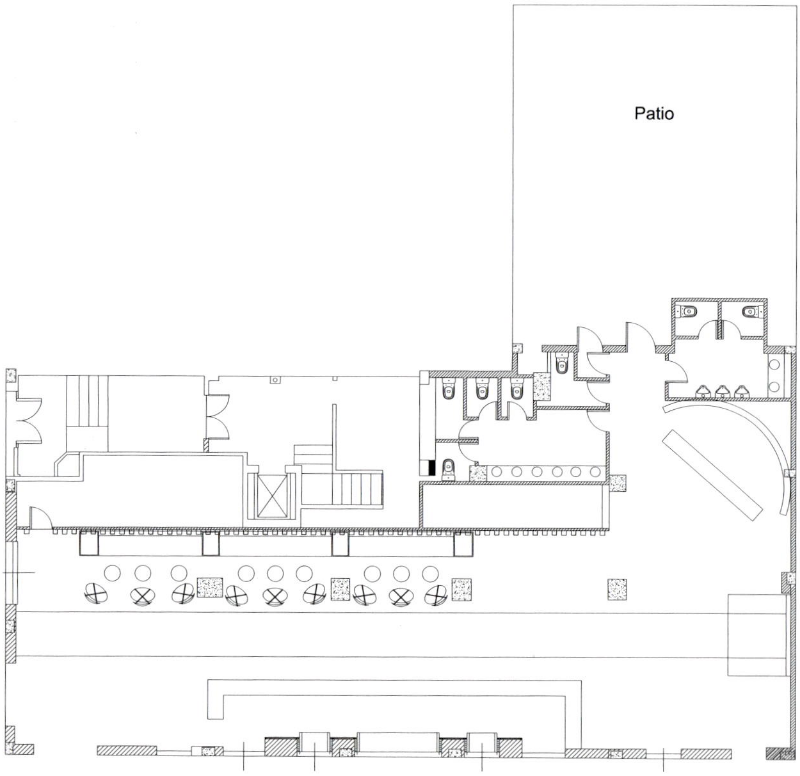

FLOOR PLAN

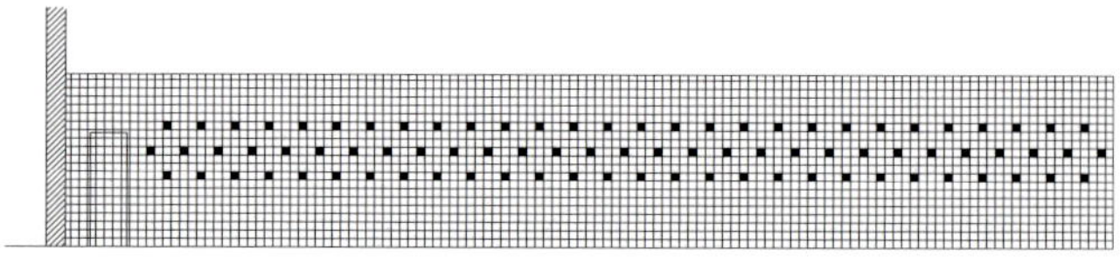

ELEVATION TILED WALL

ARCHITECT: Tomás Alía / **LOCATION:** c. Echegaray 18, Madrid / **CHRONOLOGY:** 2006 / **CLIENT:** Sabrabra / **BUILDER:** Empty / **LIGHTING:** Sadilsa / **BUILT AREA:** 240 m²

CERAMIC: Tiled wall of main space covered in matte white tiles digitally printed with bullfighters' backs

98

Sergi Serrat,
Marcos Catalán

Casa MJ

Caldes de Malavella, Spain

The clients of a lot full of pine trees had commissioned a small, single-storey house that could conceal the emptiness of its rooms in the absence of its owners.
The lot was long but not very wide, with its shortest sides facing streets with a difference in height of 3 metres.
The building was designed as if it were a sculpture; as if it were an object that had to respond to the demands of the client as well as to the requirements of the land.
This object was conceived as a box. it is set on the sloping part of the lot and is projected out towards the street. In this way, the car is parked underneath. the entrance is from the middle of the floor, both from the garage of following the ramp up from the street. The building is completely glassed in with porches on its east and west faces where there are views; it is closed facing north and south where they are nearest to neighbouring houses.
All floors, interiors and porches, make use of large (60 x 40 cm) staggered porcelain stoneware tiles. To give continuity and coherence in textures, the same tile is used in bathroom facings, but with a change in the way they have been hung – as a grid.
There is a courtyard, although this is more like an incursion by the garden into the house, which acts as a distributor separating the areas of habitual use from those occasionally occupied at weekends.
The courtyard creates cross views from the different parts of the house, traps light from the south to allow it into the living areas and entrance, and affords tangential views towards the side of the lot so that there are no direct views over neighbouring buildings.

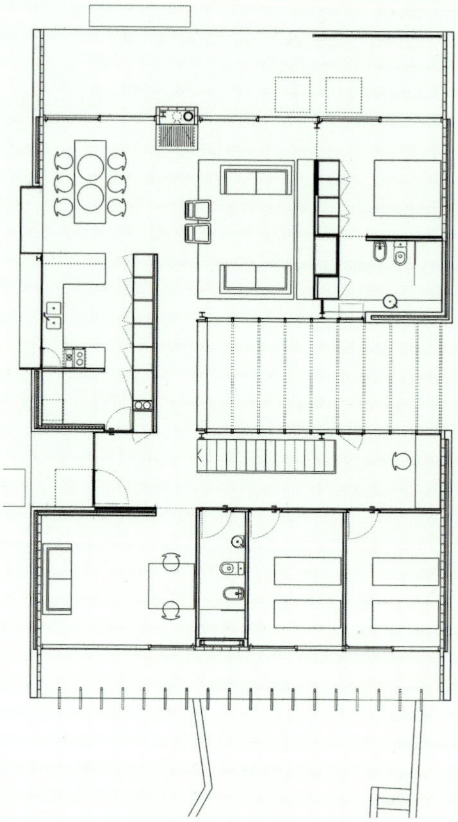

FLOOR PLAN DWELLING

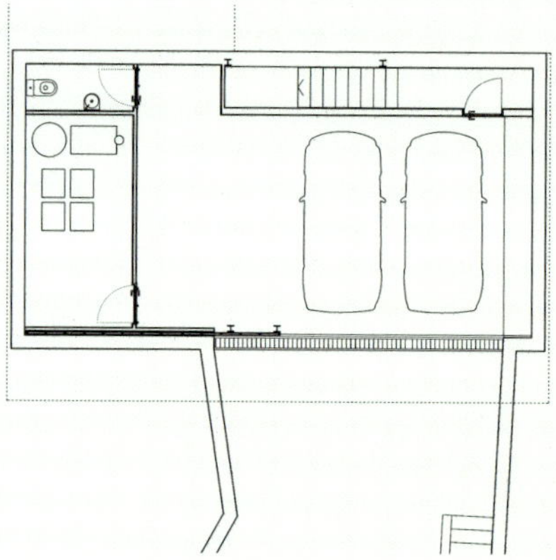

FLOOR PLAN GARAGE

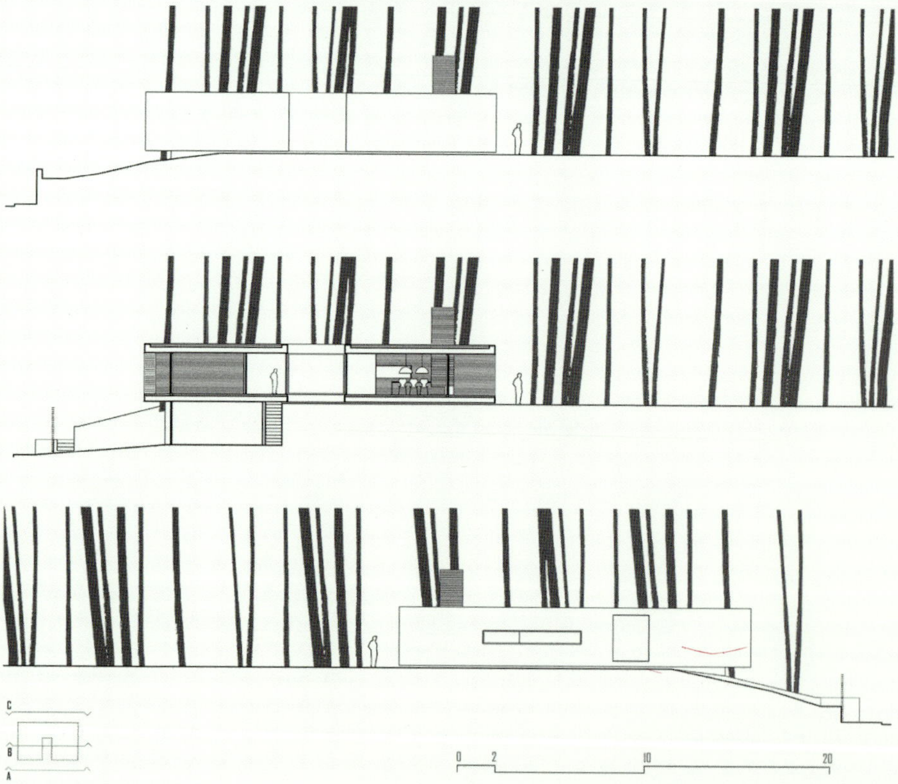

A-B-C SECTIONS

CERAMIC: Interior and porch paving: rectified porcelain stoneware tiles 60 x 40 cm laid in running bond pattern bathroom wall tiling: same as floor but laid in vertical grid

ARCHITECTS: Sergi Serrat, Marcos Catalán / **LOCATION:** Caldes de Malavella, Girona / **PROJECT:** 2003 / **CONSTRUCTION:** 2005 / **CLIENT:** Juan Lopez / **BUILDER:** Cristobal Infante / **QUANTITY SURVEYOR:** Mireia Pujades / **FRAME:** Eduard Doce / **INSTALLATIONS:** Instalaciones FECA / **BUILT AREA:** 250 m² / **PHOTOGRAPHERS:** Eugeni Pons

90

José Ignacio Linazasoro

Residential building in Lasesarre

Barakaldo, Spain

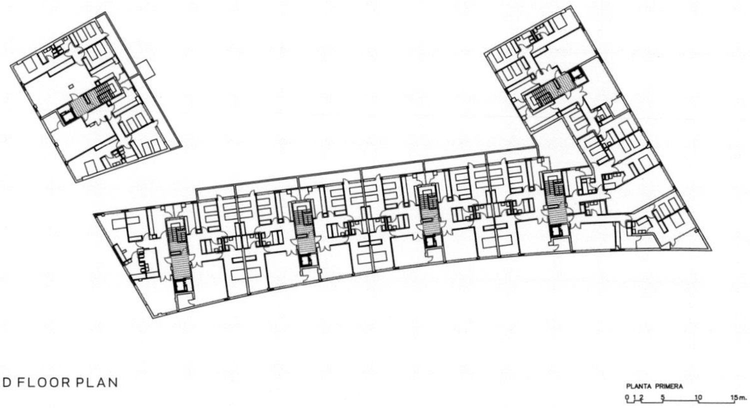

STANDARD FLOOR PLAN

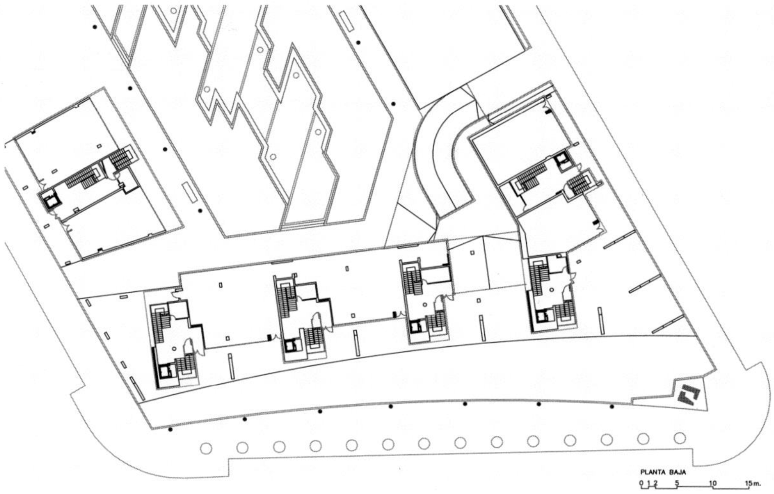

GROUND FLOOR

This project occupies half of a block shared with two other buildings.
Council planning ordinances were modified to adapt them to the project design. This involved a curved façade that created a larger area of pavement. The pavement also penetrates the building and joins on to the large courtyard at the centre of the block, turning it into a public square.
The ground floor consists of commercial premises arranged in the porticoes leading to the square.
The building is a series of horizontal bands with continuous inset lines where windows alternate with balconies. The top floor consists of dwellings with terraces that are broken up by "boxes", giving a sense of rhythm to an otherwise excessively horizontal block.
The conventional floor plan provides two dwellings per landing. The top floor has spacious roof terraces, in particular that of the sharpest corner of the building. The structure is made from reinforced concrete with V-shaped columns on the ground floor to reduce the need for many while at the same time giving a sense of expression to the building's weight as it is supported by them.

The project seeks the contrast between heaviness and lightness, the former marked by the horizontal lines and its relation with the supporting pillars and the latter marked by the continuous inset bands. Given this nature, materials were considered that would enhance these effects.
Consequently, different alternatives were considered, always based on materiality. The first idea put forward was prefabricated concrete but this was ruled out because of its uniformity. A tiled ventilated façade was then put forward and it was decided that the tiles would be of different types to give different textures when combined.
This was appropriate for the effect of variation and textured shading that was originally conceived but now enriched. The tiles come in both matte and shiny, always in a dark grey tones, with the matte tiles alternatively laid on the reverse side to enhance the texture.
The façade is finished with dark wood work and Prodema panels for the shutter housing box.
These boxes protrude slightly from the remaining woodwork to produce a horizontal line that reinforces the lines of the façade.

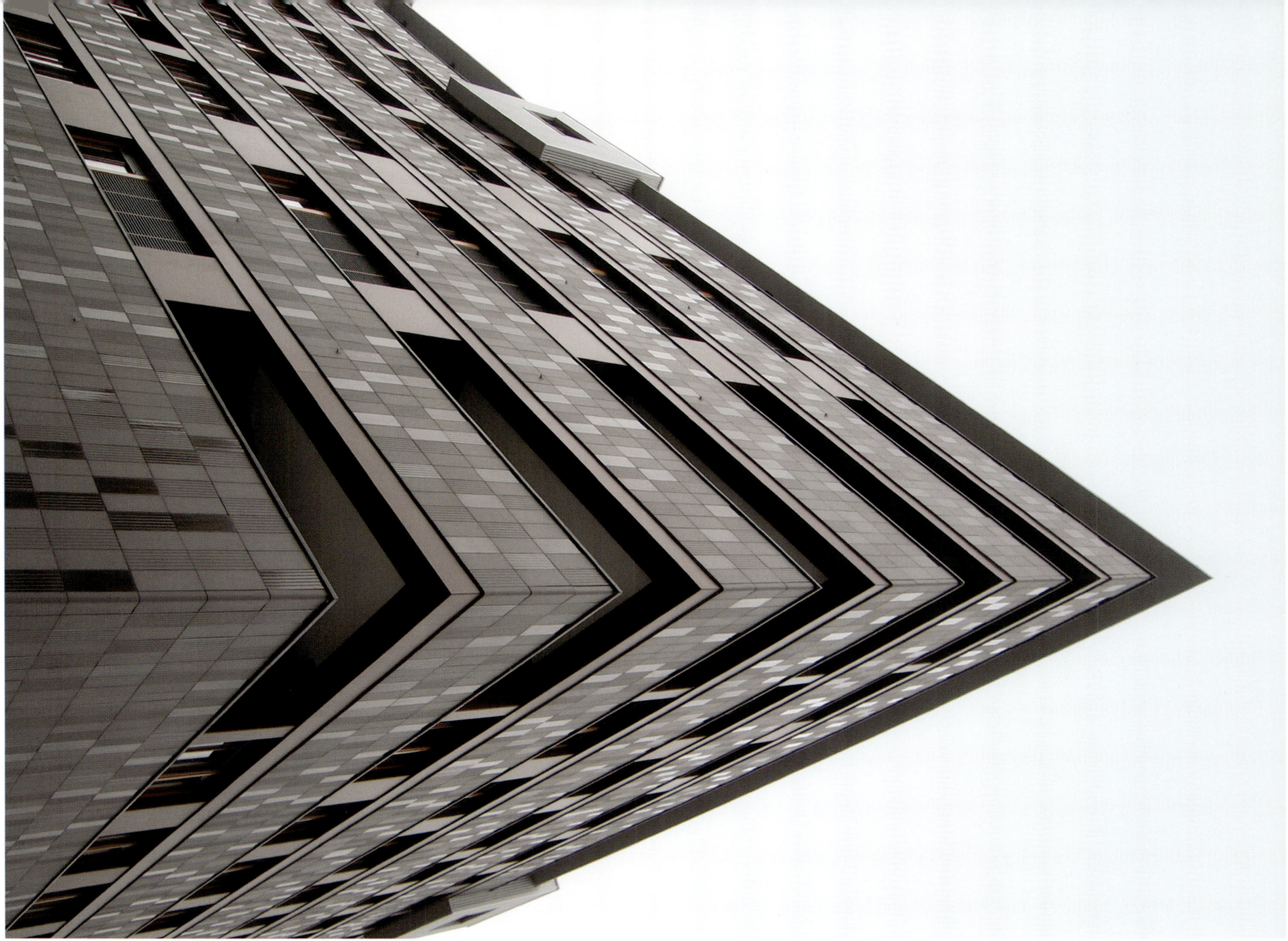

ARCHITECT : José Ignacio Linazasoro / **LOCATION** : Lasesarre, Barakaldo / **PROJECT** : 2003-2005 / **CONSTRUCTION** : 2005-2007 / **COLLABORATORS** : Hugo Sebastián de Erice / **CLIENT** : Vizcaína de Edificaciones, s.a. / **STRUCTURE** : Planexa / **PROJECT MANAGERS** : José Ignacio Linazasoro, Iñigo Saloña, Hugo Sebastián de Erice / **QUANTITY SURVEYOR** : Luis Ballesteros Suso

CERAMIC : Extruded black stoneware tiles colour applied to the tile mass dimensions 24,5 x 49,5 x 1,5 cm; some tiles with clear shine glaze fired at high temperature (1.250°C) two types of tile, one flat and one with raised textured relief some of each type glazed so that four different tile types are combined at random to give an aspect of both random vibration and constant density. Four different angular tiles for corners between adjoining façades

Basilio Tobías Pintre

Sports Hall at the Jaume I University

Castellón, Spain

The size of the plot, its location on the university campus and the conditions created by the proposed uses it would have, determined the layout and design of the Sports Hall, with a large ground level area housing most of the activities.
The building is set out in three longitudinal bands of differing heights to accommodate the different uses it would receive. Most of the building area corresponds to the multi-purpose hall, with interior dimensions of 88.5 by 30 m and a free height of 12.6 m. The central band has three levels, while the north-east facing band is two or three levels in height depending on the areas.
A long band of storage areas divides the multi-purpose hall from the central band. At ground level, there are the different changing rooms and a wide space for administration and user services. The different halls and gymnasiums are located on the first level. Two classrooms, one for observation, are situated over the offices and are connected to the multi-purpose hall. Different service installations for the building are located at either end of the roof level of this central band leaving an outdoor space that can be used for sports activities.

Special attention has been paid to allowing independent traffic flow for the different users of the facility.
The outline and the construction system adopted were decisive in the building's layout and in giving it a unified structure. The short execution time scale and the optimisation of building processes led to the use of a metal frame. This is visible from the outside and expresses the logic used in construction.
A large proportion of the opaque infill panels on the façade are tiles, attached to the metal frame to create a curtain wall onto which elements created from traditional material have been integrated. By means of construction and assembly using precision technology, this has achieved a façade with excellent thermal and acoustic properties, while its colour allows the building to blend into the setting created by the other buildings on the campus. The openings left in the external panels are protected by electronically-controlled vertical metal louvres that allow the sunlight entering the building to be regulated while affording protection at the same time.
The use of faceted metal for the louvres, integrated into the structural reticulum, confers the image of a changing membrane to most of the open wall spaces of the hall.
The two inner courtyards play an important role as certain meeting areas and passageways open onto them. The courtyard located on the ground level is a green space and the cafeteria opens onto it whereas the main one found on the second level is more of a roof terrace as it can be opened up to the outside by means of louvres.

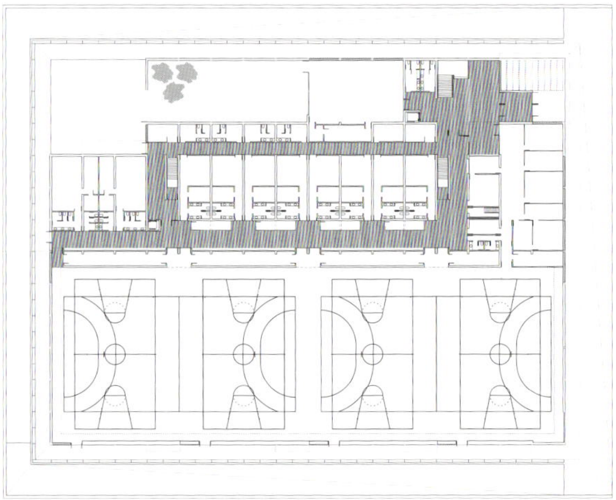

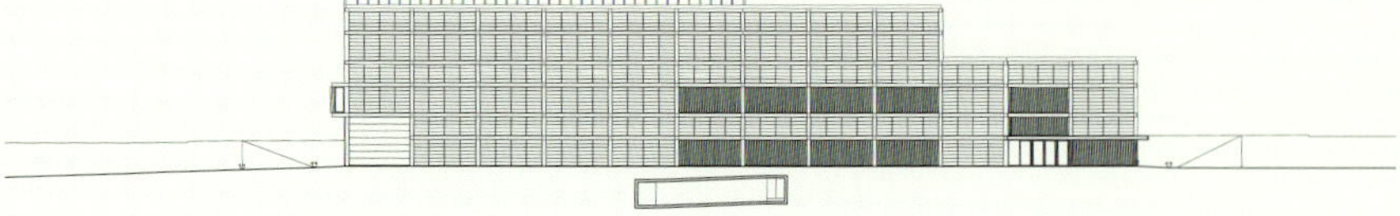

SOUTH-EAST ELEVATION

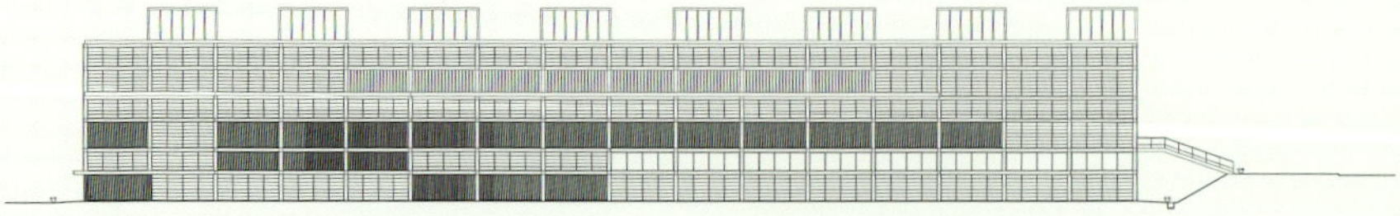

GROUND LEVEL

NORTH-EAST ELEVATION

SOUTH-EAST ELEVATION

NORTH-EAST ELEVATION

CERAMIC: Curtain wall formed by opaque infill panels on elevation 127 x 30 cm shingle-type tiles attached to galvanised aluminium T profiles with stainless steel staples

ARCHITECT: Basilio Tobías Pintre / **LOCATION**: Parcela nº 15. Campus del Riu Sec. Universitat Jaume I. Castellón / **PROJECT**: Noviembre de 1999 / **CONSTRUCTION**: Julio de 2002 / **COLLABORATORS**: Santiago Carroquino, Jaime Magén / **CLIENT**: Universitat Jaume I. Castellón / **QUANTITY SURVEYOR**: Enrique Morales Simón / **INSTALLATION CALCULATIONS**: Pedro Bellido. Inco Ingenieros Consultores / **FOUNDATIONS CALCULATIONS**: Ateco / **STRUCTURE CALCULATIONS**: Ingemetal / **BUILDER**: Constructora San José s.a.

96

Francesc Rifé

Offices Cava Recaredo

Sant Sadurní d'Anoia, Spain

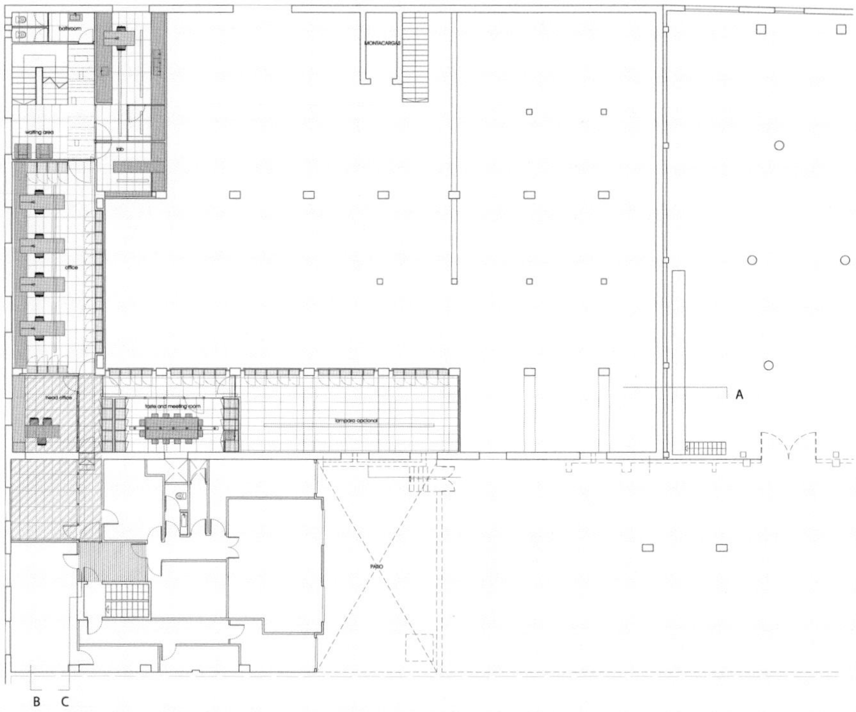

FLOOR PLAN OFFICE LEVEL

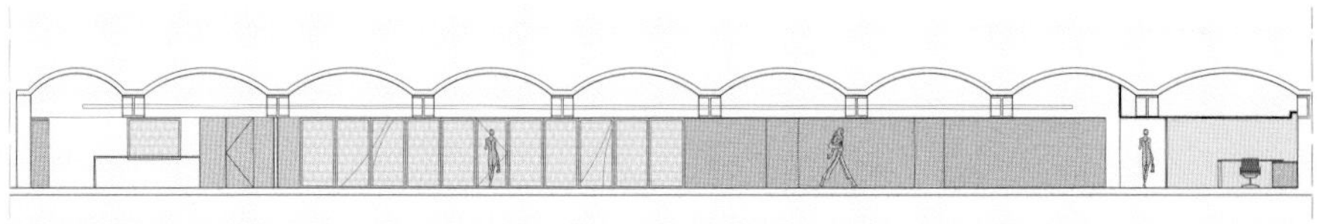

SECTION A-A'

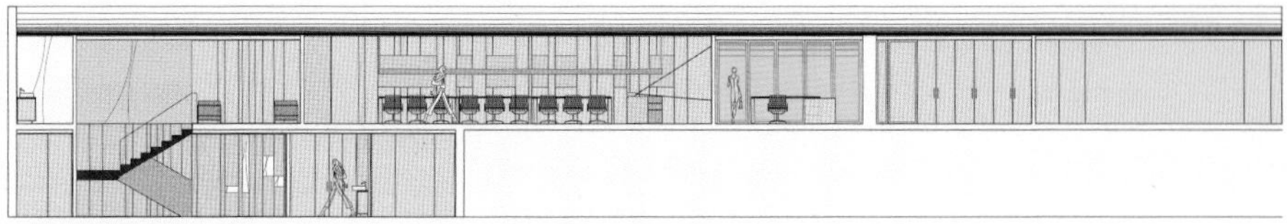

SECTION B-B'

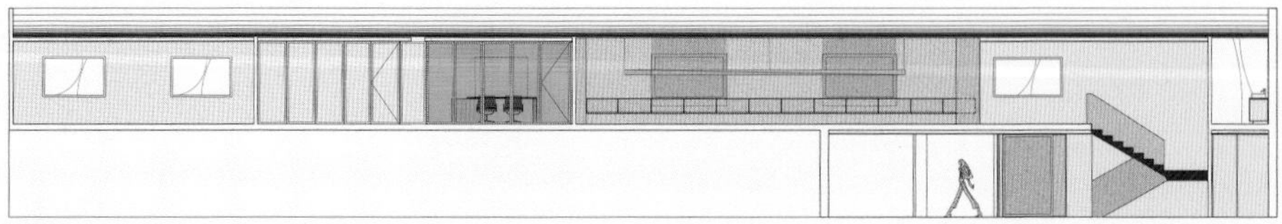

SECTION C-C'

The starting point for the project was a warehouse adjacent to the original offices and cellars, a two-storey complex with a dilapidated façade and deteriorating brick on the inside.
In order to maintain coherence with the surroundings, it was made clear that the building and its materials should be kept practically the same.
As a result of the design, the visitor enters a small black-painted brick vestibule with a metal staircase. Once upstairs, there is a waiting room, a bathroom and a laboratory.
The floors are covered in brown, specially-designed ceramic tiles. As the surface is slightly indented lengthways, the floor is non-slip and has an elegant texture. This also makes the joints between the tiles seem to disappear. The tile is perfect for the standards of cleanliness required in the building, which is why it is also used on the bathroom wall.
For the Ricaredo design, the dark brown and light brown are complementary and are used in such a way as to create a path, one that guides the visitor can use to move around the building.
Continuing on, there is the office, also a very aseptic area decorated basically in white, and a meeting room that is also used for wine tasting.
A basic element of the design was the ceiling. The curved frame was renovated and reinforced. It was painted white to enhance the natural light entering the building and to contrast with the darker colours and the metal beams. The shape of the ceilings made it impossible to install flush lighting, so a special indirect lighting system was designed.

ARCHITECT: Francesc Rifé / **LOCATION:** Sant Sadurní d'Anoia, Barcelona / **CHRONOLOGY:** 2005 / **CLIENT:** Ton Mata / **BUILDER:** Blausmo Santa Coloma, s.l. / **INSTALLATIONS:** Simac Instal·lacions, s.l. / **ELECTRICITY AND LIGHTING:** CROMM Il·luminació, s.l. / **CLIMATE CONTROL:** Tecnotrol, s.l. / **BRICKWORK, TILING, PLASTERING:** Blausmo Santa Coloma, s.l. / **METALWORK:** Demetall / **WOODWORK:** Denia, s.l., Bonanova, s.l. / **PAINTING:** Pintura i Decoració Murtra, s.l. / **GLAZING:** Cristalería Bonanova, s.l. / **BUILT AREA:** 380 m² / **PHOTOGRAPHER:** Eugeni Pons

CERAMIC: Floor tiles 90 x 45 cm brown designed by Francesc Rifé

Ceramic Studies Departments

In its drive to broaden knowledge of ceramics and their applications in professional spheres, ASCER has set up a number of Ceramic Studies Departments in different architecture schools in Spain. The network of these departments puts into practice the involvement of the University in the ceramic tile manufacturing sector; this involvement is understood as the mutual exchange of knowledge and experiences. On the one hand, the Departments enable the tile industry to add greater understanding of techniques regarding this product to the training of future architects, in addition to the huge aesthetic possibilities that ceramic tiles can contribute to the creative work of this group of professionals. On the other hand, it allows future architects to direct their concepts towards ceramic use by innovating already-existing formats or by developing new applications.

The network of Ceramic Studies Departments comprises:

—Ceramic Studies Department at ESARQ (School of Architecture) at the International University of Catalonia.

—Ceramic Studies Department at ETSAA (School of Architecture) at the University of Alicante.

—Ceramic Studies Department at ETSAV (School of Architecture) at the Polytechnic University of Valencia.

—Ceramic Studies Department at ETSAM (School of Architecture) at the University of Madrid.

—Ceramic Studies Workshop at the Jaume I University in Castellón, as part of their Architectural Technology degree.

The educational methodology of these Departments is based on Workshop teaching, a place of ongoing work and meeting point for teaching staff and students where debate, reflection and research are encouraged.

Different academic activities are held by the Departments, among which special mention can be made of: technical seminars, visits to factories and buildings of particular interest owing to their use of ceramic materials, lectures given by renowned and prestigious architects and professionals from the ceramic sector, round table discussions, etc. There are also complementary activities such as visits to innovation centres (e.g. ITC), R&D programmes for specific initiatives by companies in the sector, design competitions for student projects, project exhibitions (Expocátedras), publication of reports detailing the results of the activities undertaken.

Víctor Echarri

Ceramics and systems installations between integration and design

The Ceramic Studies Department at the Alicante School of Architecture

Architecture is facing complex challenges. Never before have such unique projects been undertaken with such sophisticated materials, genuine outpourings of energy, buildings that defy the imagination but justified only by the uncertain thesis that is progress.
But there is another reality that coexists with this. People who have to ration their scarce resources in an attempt to provide decent environments where they can develop themselves as persons. We proudly speak of sustainability when it is only on limited occasions that we use our reason to adapt ourselves to a common good, to what really matters in a world in need of solidarity.
The building industry is steaming ahead towards never-imagined quality, towards new materials with dream-like performance, among which ceramic undeniably holds a place of honour. Research in the ceramic sector has been a source of great satisfaction in recent decades. We are on the threshold of a time when the use of ceramic will be extended to those areas of construction occupied by other building materials. Only imagination, study, work and rationality are needed. Rationality is important because it is necessary to have in-depth knowledge of current manufacturing processes, their limitations and their possibilities for application. Rationality is essential because , in addition to the excellent results achieved in unique buildings such as the Santa Caterina Market in Barcelona and the Spain Pavilions for Expos in Aichi and Zaragoza, ceramic still has to compete with other materials in fields as different as furniture, service installations, dropped ceilings, raised floors, etc.
I think that technological advances have been incorporated to a large degree in building installations. In recent decades, domotic automation systems of different kinds, new climate control systems, telecommunications installations and a long list of systems that need integrated solutions, giving a larger number of users access to these advances. The world of installations has gained in importance and we take increasing notice of what is going on inside in a bid to surpass their performance and to accommodate their presence in buildings as intelligently as possible.. It is no wonder then that this is the direction taken by the Ceramic Studies Department in Alicante in its first steps after being established. A panorama is opening up to us that is undeniably full of opportunities to continue to dream of the presence of an excellent material – ceramic – in those parts of a building that have traditionally been entrusted to other materials.
The ceramic market is currently involved in innovation that is almost exclusively based on imitating other materials. However, competition within the sector, may mean that this idea is starting to run out of steam. The general feeling seems to be that it is necessary to move away from imitation and seek something more genuine, to give ceramic a more active presence in the formal and aesthetic structure of our buildings. The challenge taken up by the Ceramic Studies Department at the University of Alicante, based on the exploration of the integration of ceramics into installation systems, seems to fit perfectly with the general preoccupations of the sector. These pages display some of the projects presented and awarded prizes by the respective juries and which are examples of our commitment to new horizons discerned by ceramic.

Manhole covers

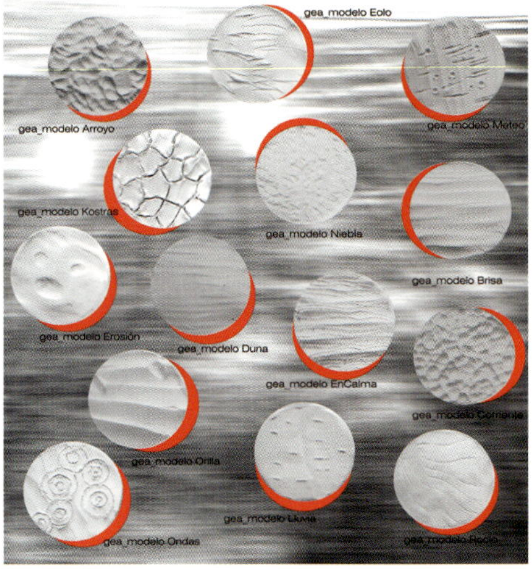

Urban infrastructure has need of many manholes for proper execution and subsequent operation. It would surprise you to know that in barely one kilometre of city street, there are around 300 manhole covers for sewerage, telephone, telecommunications, water supply, natural gas and other utilities. Owing to its strength and price, the universally accepted material is cast iron. Currently, research work is being carried out focused on improving the rheological properties of ceramic in order for it to acquire sufficient resistance to impacts and strain caused by heavy traffic - ductile ceramic.

The idea developed for ceramic manhole covers comes under the framework of a future urban landscape. Over the last few years, the use of non-slip porcelain stoneware has been significantly developed for public spaces but the incorporation of manhole covers in the same material has not come about. This would enable total continuity to be possible when planning urban landscaping with all kinds of textures, colours, raised patterns and finish styles – matte, shiny, etc.

The manhole cover is designed as a combination of three layers. The middle layer is a core of light, tramex-type steel which would give sufficient strength to the piece until ductile ceramic is a reality. The upper and lower level are of porcelain stoneware, reinforced with fibreglass mesh and attached to the steel core. This allows the system to be reversible so that it could have a specific design for one part of the year and another when the manhole cover is turned over.

This project is the first patent developed at the Ceramic Studies Department in Alicante and we expect it to be at the prototype stage soon.

101

Triangular tiles for raised paving

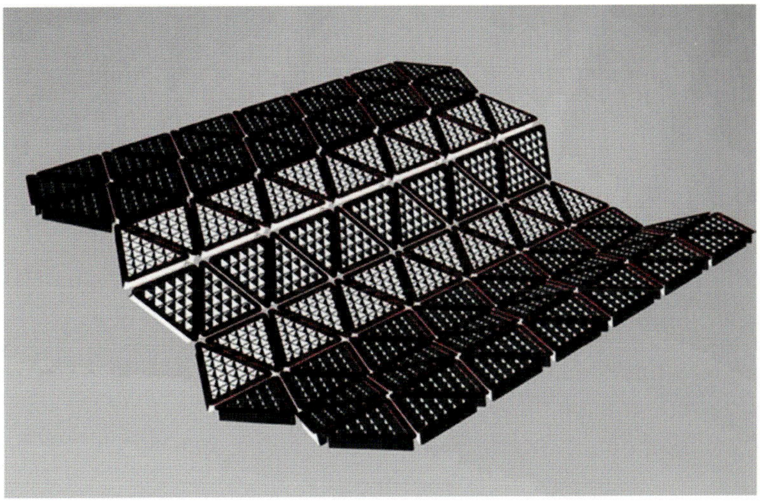

The concealing and access to systems installations in the roofs of buildings was the inspiration for the development of a triangular tile, in turn made up of two triangular tiles. The base is a more economical, extruded stoneware tile while the finish is in pressed porcelain stoneware, attached to the base with epoxy resin. The tile is designed to be anchored onto a metal frame with standard profiles such as IPE, IPN o HEB. This allows total access to the systems installations. The search for or exploration of finishes was based on the degree of opaqueness through a higher or lower number of also triangular perforations allowing the development of microtopographies of vegetable covers or combinations with other materials. An interesting area for action would be the study for day-night reversibility with the integration of LED lighting for use in advertising, communication, etc.

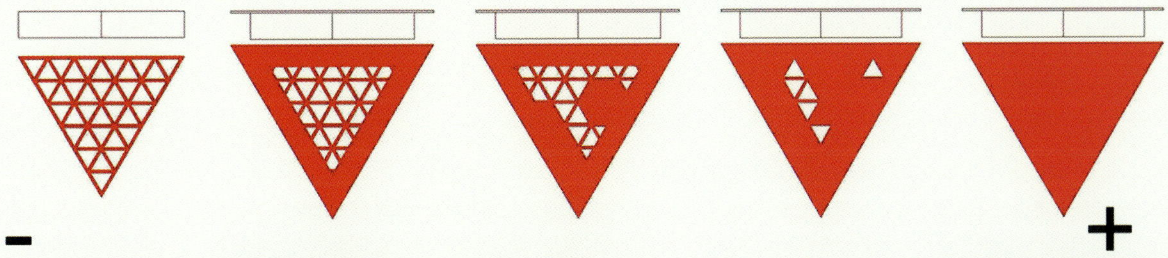

Rotatri
Triaxial tile for lattice-like spatial compositions

The project, given the name of ROTATRI by its creators, does not intend to conceal the systems installations that can run along the façades of buildings, but rather mimics them by highlighting the ceramic tiles contrasting with the setting. The tile is made up of three parts: a central core and three linear parts each at an angle of 120°, 15 cm long.
The linear parts have slots or linear indentations that allow them to be overlapped with others to form flat or three dimensional lattices. This three-dimensional structure is achieved by placing the slots on the same plane as the tile or at right angles, lengthways or perpendicular.

The study of scale is interesting.
It is present in the combination of tiles, from the simplest, most specific use on a façade to the configuration of an urban complex – from a simple abstract enclosure to a full installation. The mimicry achieved allows installations to be understood as something that should not be hidden or as unnecessary for the urban image, but as necessary elements that speak and explain how buildings live, how they feed, how they are formed. This organic image is enhanced by the tree-like image conveyed by the ceramic tile.

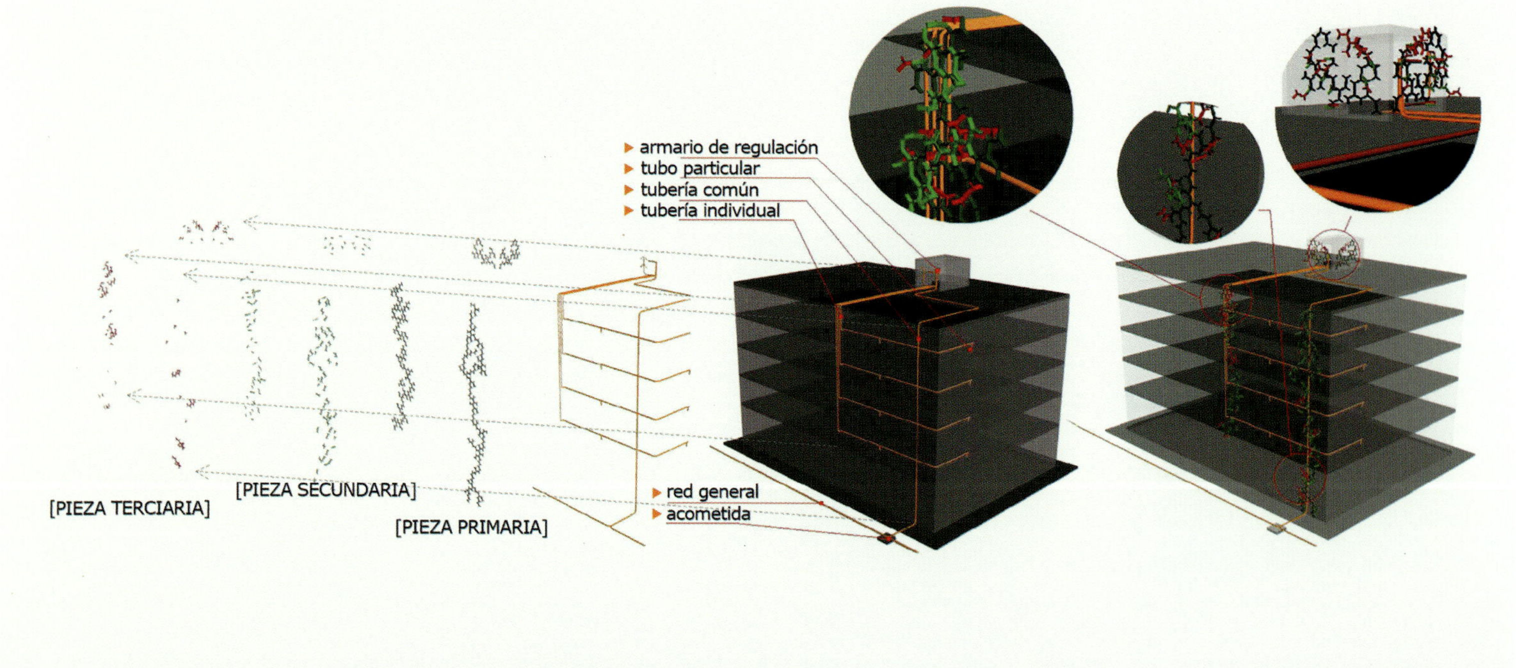

103

Press Button project

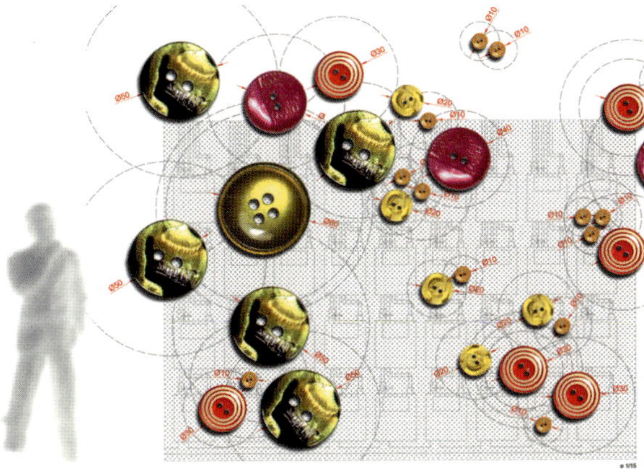

Pop art graphics is the base for the design of circular ceramic tiles anchored on perforated metal plates. A building system is developed with metal-framed modules measuring 1 x 1 m, strengthened by a diagonal bar and covered with perforated metal. This module can be used to form skins and, in particular, as a solution to the ventilation and accessibility of central meter cabinets, the subject proposed for this exercise. The novelty of the system is in how the described module is made with the incorporation of circular ceramic tiles of different diameter, anchored onto the metal plate. In this way images can be made to cover large surfaces.

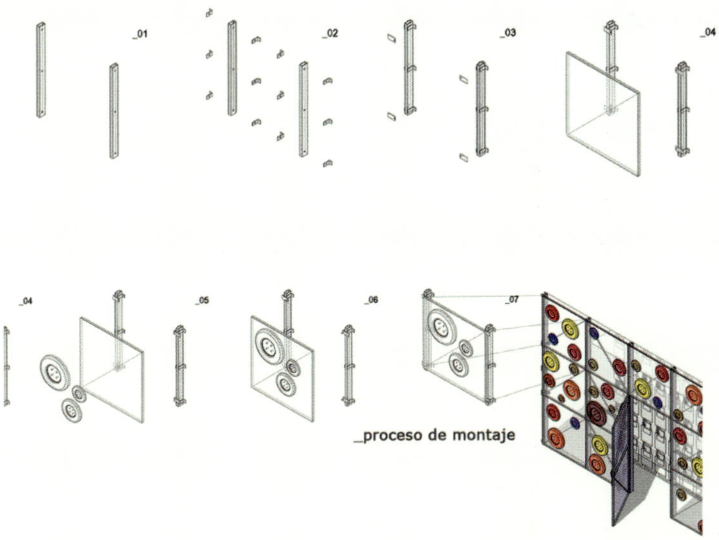

104

Threaded ceramic tubes

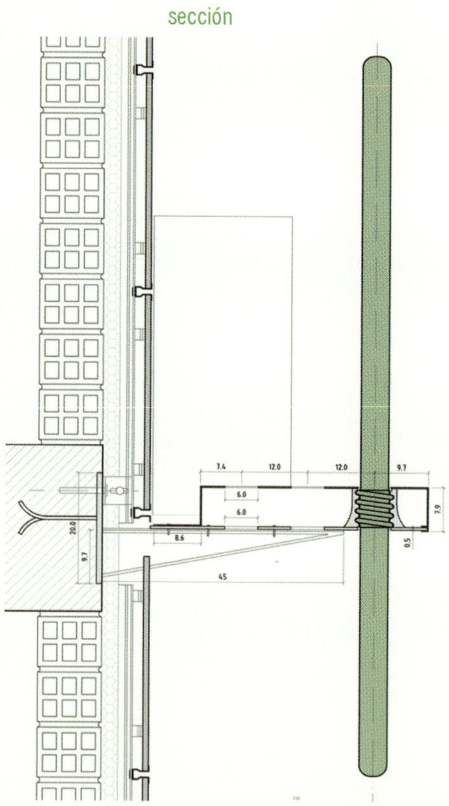

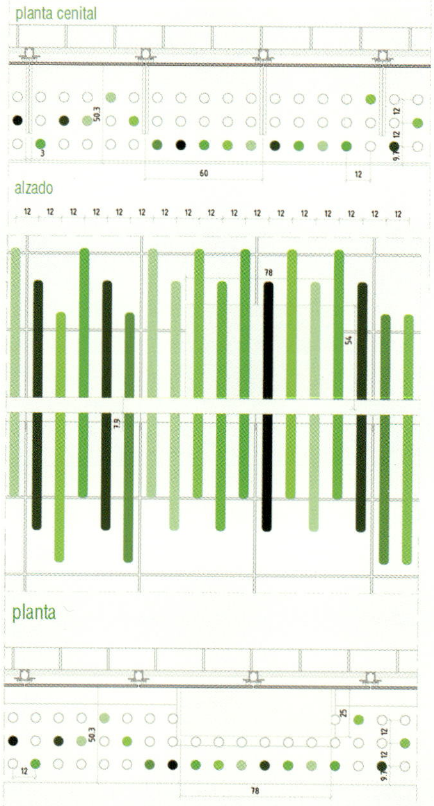

This is the first of the projects undertaken to find a way of integrating air conditioning elements through the use of porcelain stoneware. The objective of the project is to solve the aesthetic problems caused by the usual positioning of external units in domestic air conditioning systems. The solution proposed consists of an anchoring system for threaded ceramic tubes screwed on to a metal platform running the length of the façade at the floor level of each storey. The external units are placed on the platform. The ceramic tubes, made of porcelain stoneware, are 130 cm long, although these may vary. The thread allows the tube to be screwed into the platform. Two different models of tubes can give a vibrant aspect to the composition. One of them would have the thread halfway along, while the other would have it one third of the way. The system would allow the density of tubes to be varied and, with multiple textures and colours, they could be adapted to practically all combinations and fancies of colour as desired by means of pixelation. The construction process is carried out in three differentiated stages.

105

DIAFRAMICA
air vents

We have always thought of air vents as circular and static elements, of better or worse design, but always subject to simple functionality. The use of porcelain stoneware for the manufacture of air vents seemed a viable alternative; owing to the scarce absorption of moisture, it is a particularly hygienic material for air conditioning.
The idea went far beyond that. It was an attempt to escape from the conventional funnel shape – a static and passive element – and it was proposed as the protagonist of the installation, to be the very expression of how the system works. The vent was designed as a living being, based on opening flowers. Six trapezoidal tiles of varying thicknesses interlock so that when the system has no pressure – no air is released – the unit remains closed to form a hexagon. When the fan introduces static and dynamic pressure into the system, the six pieces move and appear as a hexagonal aperture. Internal springs make it close when the fan stops and there is no longer any pressure. Thermochromic glazes are applied to indicate if the air entering is cold or hot.

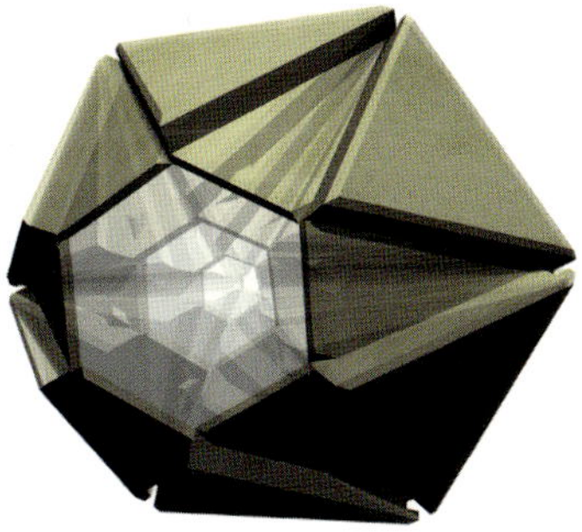

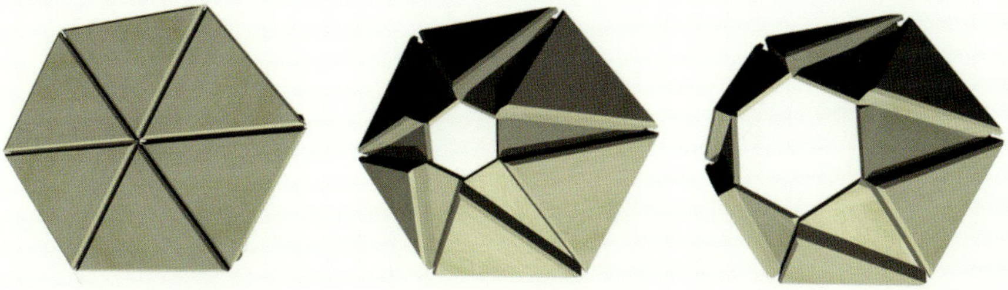

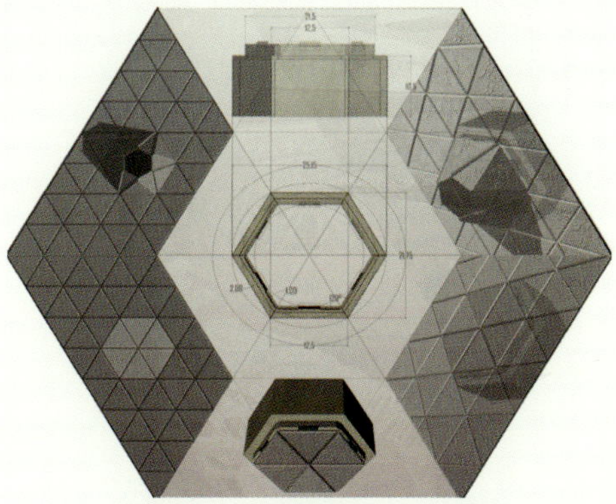

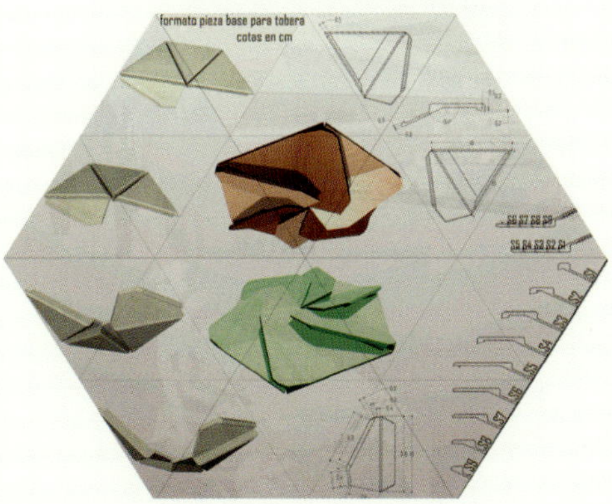

CLIM(b)ER façade

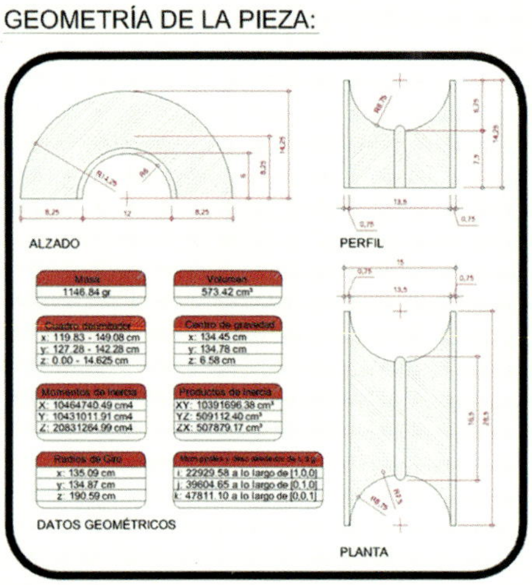

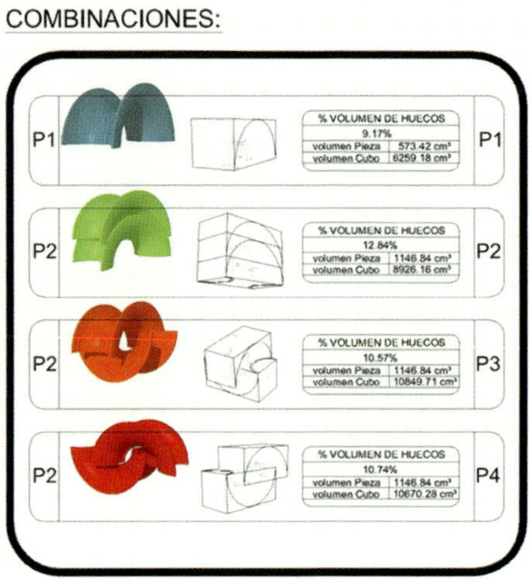

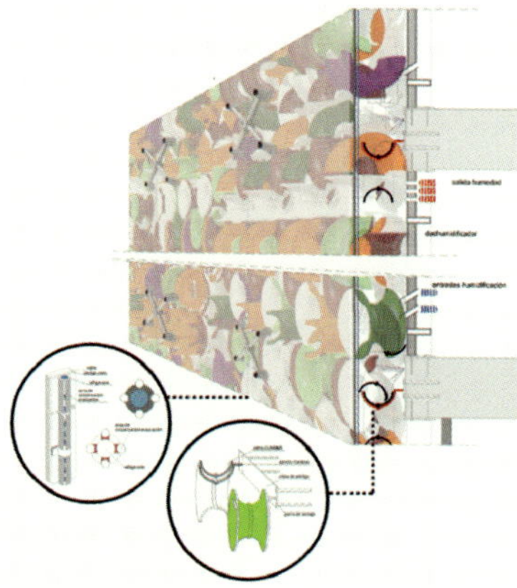

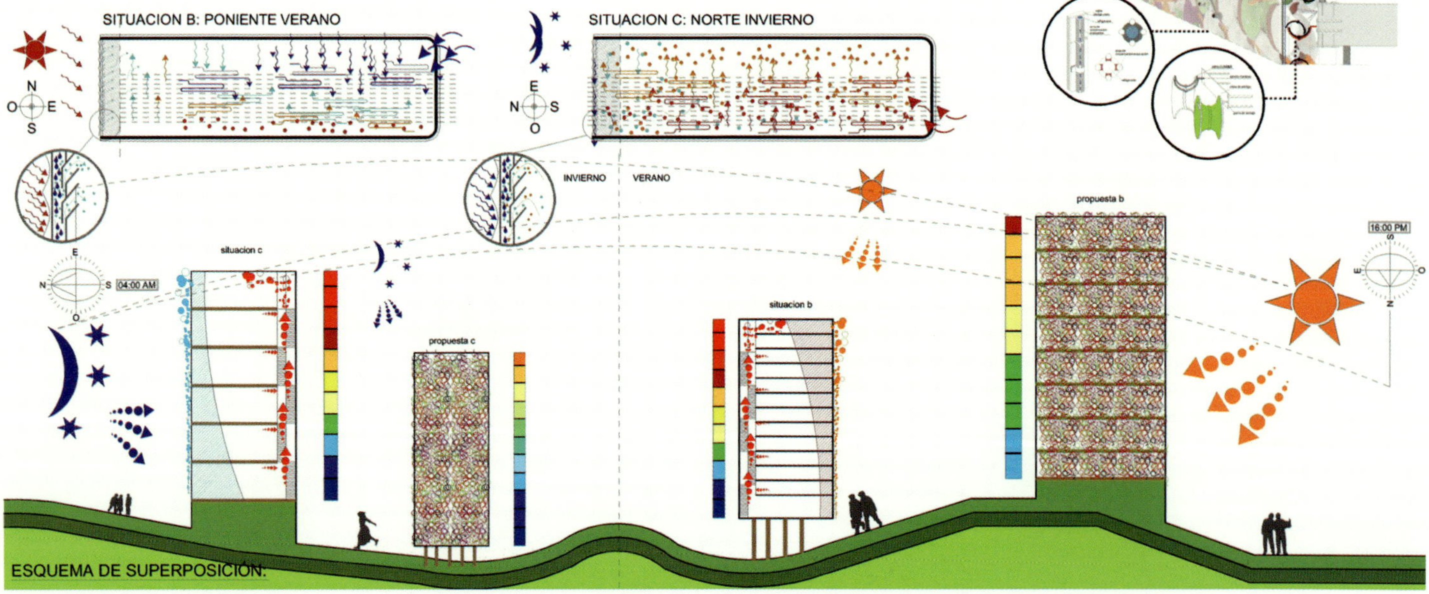

Based on the distillation tower effect, where contact by the passing of airflows causes a heat exchange between the descending cold air masses and the very ascending hot air, this system incorporates façade surfaces comprising two glass planes packed with toroidal ceramic tiles. These assist the sustainable development of architecture through three possible climate control effects produced with low energy use.

1. Packing. This consists of ceramic tiles with maximum surface area contact with air and moisture. They have the largest number of openings to avoid condensation occurring in the inside of the chamber. Hot air masses rise at lower speed and temperature.

2. Humidifying concrete slabs. These circulate water that is evaporated from the faces of the slab countering the effect of solar radiation and causing the temperature of the air circulating in the chamber to descend notably.

3. Dehumidifying façade. In a situation contrary to that previously described, condensation is produced in the chamber by contact with surfaces at a lower temperature, resulting in dehumidified air that will later flow through the interior of the space, improving the air conditioning effect.

Gres-BOT

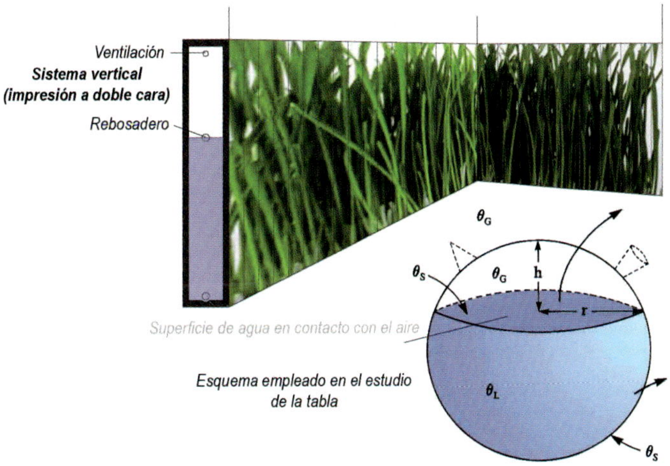

The project is based on the clay jug heat loss effect which is the effect of lowering the temperature of a body of water that has previously been introduced into a container. This container is made from extruded ceramic tiles. The system works through the effect of radiation on an internal surface that has a lower temperature than the air. The system also allows hot water to circulate in winter.

Detalle de los tubos de conexión

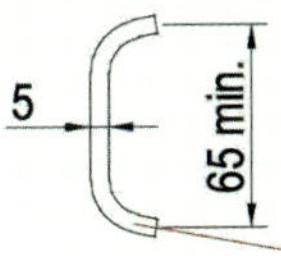

Detalle de funcionamiento del sistema horizontal
Mayor superficie de agua en contacto con el aire implica una mayor eficacia en el sistema.
Llenado y vaciado de las piezas por gravedad.

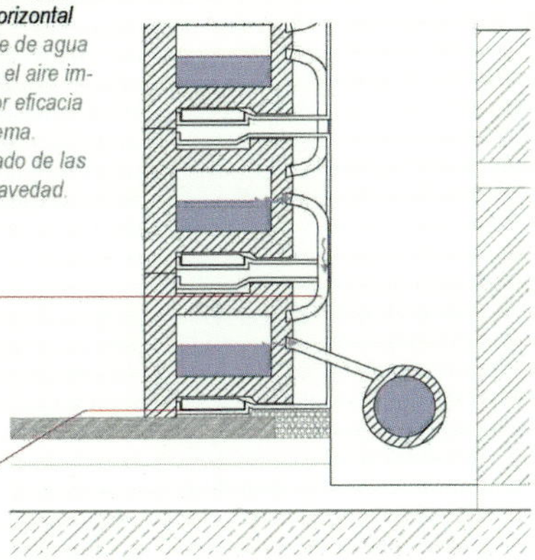

Detalle del clip de anclaje (Gres-BOT horizontal)

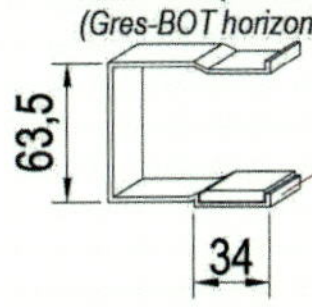

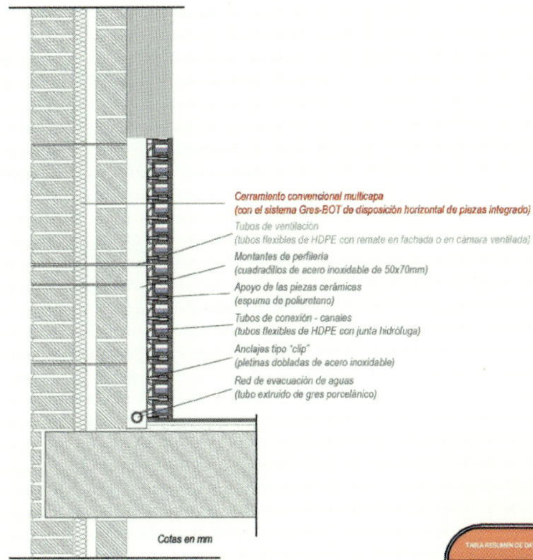

Procedimiento de montaje - Despiece

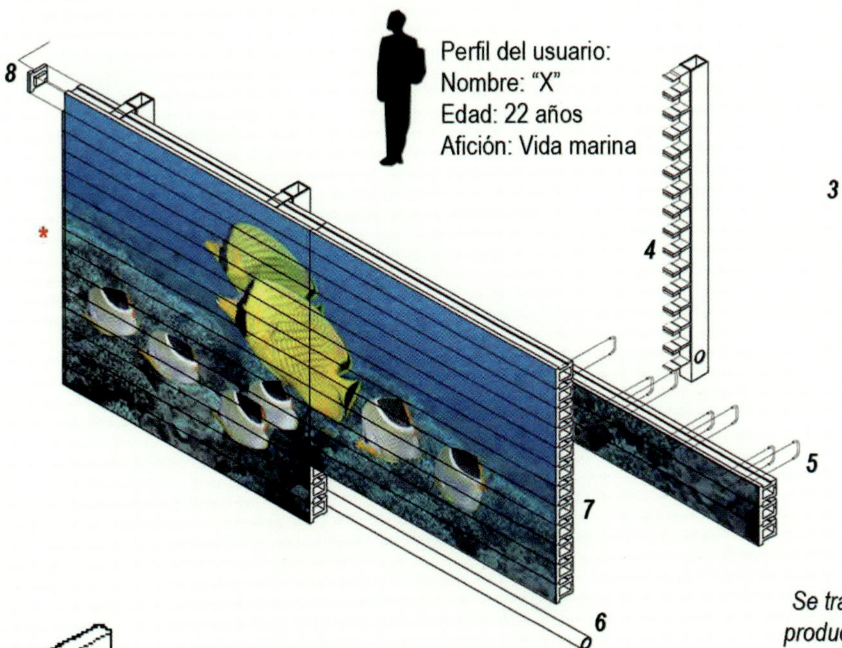

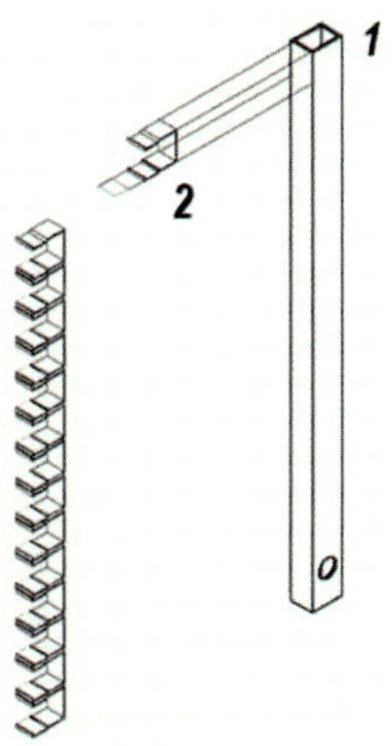

TABLA RESUMEN DE DATOS. VIABILIDAD DEL PROYECTO	
Valores Experimentales del Recinto	
Valores Experimentales en Verano	
V: Volumen o Masa de agua (g)	34.000
Cp: Capacidad Calorífica del Agua (cal/g)	1
T: Temperatura del Agua (°C)	21
hc: Coeficiente de Convección (W/m ^{2}°C)	12
a: Superficie Externa del Agua (m ^{2})	30,6
Tg: Temperatura del Aire (°C)	30
Ts: Temperatura de la Superficie del Agua (°C)	25
σ: Coef. de Radiación de calor	0,04
Superficie Total Recipiente (m ^{2})	22,4
s: Superficie del agua en contacto con el aire (m ^{2})	17
U: Coeficiente de transmisión de calor del agua [illegible]	150.000
φ: Calor de vaporización del agua (cal/g)	585
K: Coeficiente de transferencia de masa para el agua [illegible]	5
Hs: Humedad de saturación	0,8
H: Humedad del aire	0,6
Tiempo (24 h)	[illegible]

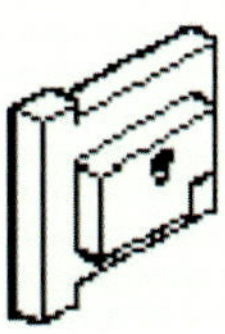

Detalle de la tapa de las piezas cerámicas/extremos (piezas cerámicas por prensado y corte hídrico)

* *Sistema "Print-In"*

Se trata de un sistema de acabado que se produce según las necesidades del usuario. Los diferentes tratamientos superficiales permiten formalizar un catálogo ("In & Out" y "Traslúcido") que permita su incorporación en el paramento de nuestra arquitectura.

Vicente Sarrablo

Trilogy

The Ceramic Studies Department at the International University of Catalonia School of Architecture

The Ceramic Studies Department at the International University of Catalonia is well aware of this important moment for ceramic tiling in architecture of great prestige, with projects that have avoided succumbing to the vulgarisation of solutions and have taken on the challenge of surpassing the achievements of their predecessors. We acknowledge that their success is being widely disseminated.

At the same time, we contemplate how the ideas of our architecture students have also made great advances in the third year that the Ceramic Studies Department has been operating. Funnily enough, when we showed students the projects of the last two years at the beginning of the year, they were initially stunned at the challenge they were facing. However, by the end of the year, four of our students had won the main prizes at the international tile design competition; Indistile, and ten projects were exhibited at FAD, the Architecture and Design Foundation in Barcelona.

Our students also rejected the easy recipes for solutions and have made the most of their achievements as an incentive for more daring designs. That has always been the spirit of architectural pioneers.

Here we show some of the most significant work from that produced in our third year. We would point out the elements in common with the previous years. All students were obliged to control their designs with great geometric precision, to offer a panoramic view of their application in architecture and to create a ceramic model of their design. All of them were also grouped into four families researching ways to transcend the standard tile format in contour (cellular tiles), profile (streamlined tiles), in their volume (faceted tiles) and density (spongiform tiles).

In the family of cellular tiles are *Rótulas* (Ball joints) by Gonzalo Martínez and *Crisálidas* (Chrysalides) by Belén Callejas, elaborate paving tiles halfway between ceramic cobble stones and boulders. Mercè Argemí and Hugo Cortés (page 112) present tiles for façades, and have designed original anchoring systems and spectacular latticework, respectively.

Rótulas (Ball joints)
Gonzalo Martínez

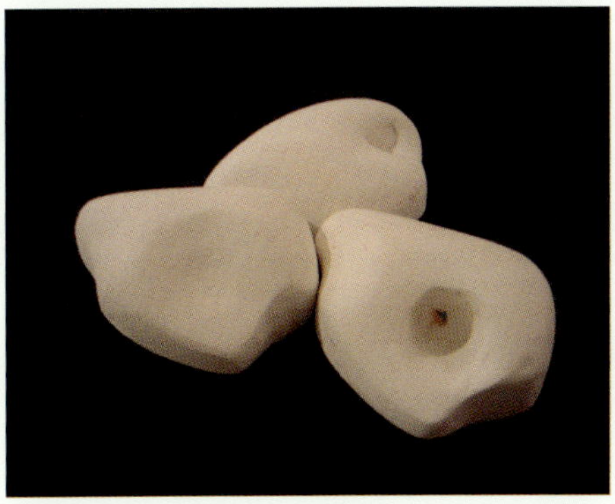

Crisálidas (Chrysalides)
Belén Callejas

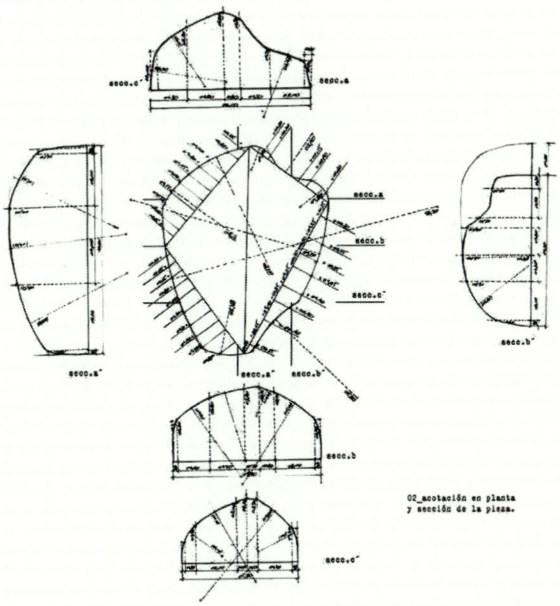

Façade anchoring system
Mercè Argemí

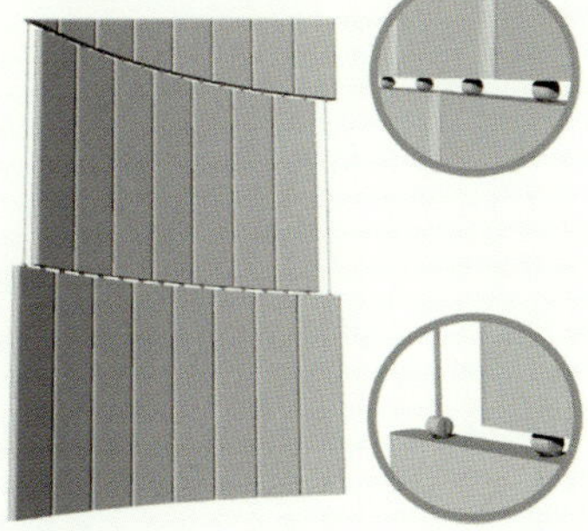

Among the streamlined tiles, *Origami* by Erika Biarnés, *Urdimbre* (Warp) by Anna Albir and *Deployé* (Unrolled) by Jordi Ayala. They are exciting textile effects for open-work walls that work as separators.

Façade anchoring system
Hugo Cortés

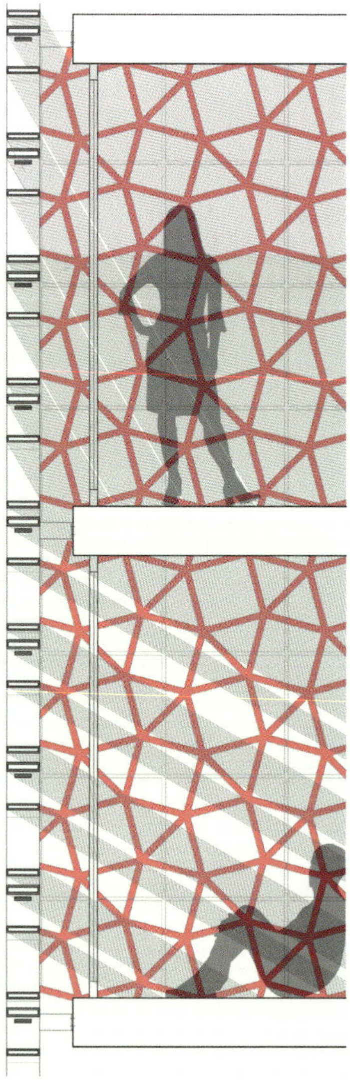

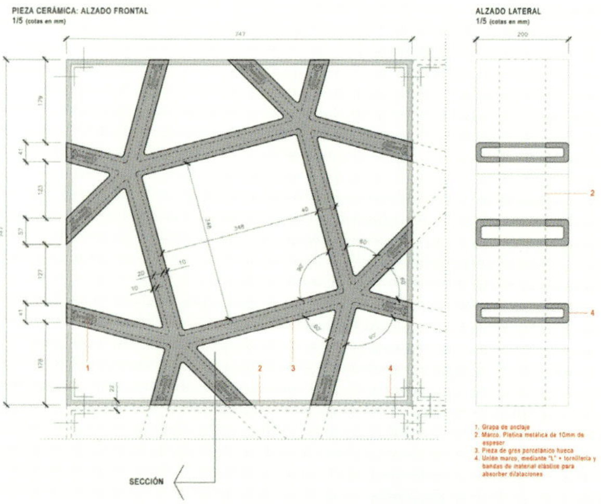

Origami
Erika Biarnés

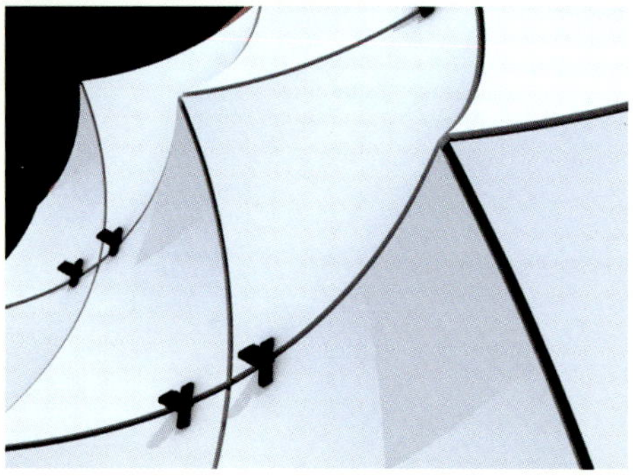

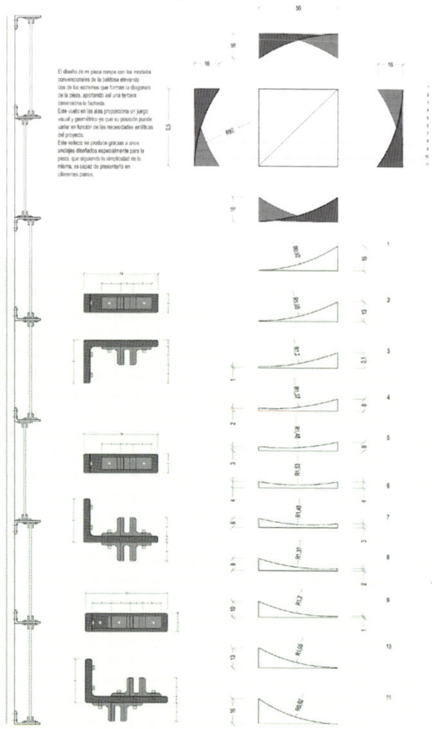

Urdimbre (Warp)
Anna Albir

Cubilote (Cupola) by Maria Roca Ribes is a faceted tile that forms curved façades thanks to its hinged fitting system. Spongiform tiles offer novel paving tile formats, such as the pressure moulded *Onda* (Wave) by Jorge Salinas and the H2orizón grille tile for swimming pools by Ester Laso (page 114).

Deployé (unrolled)
Jordi Ayala

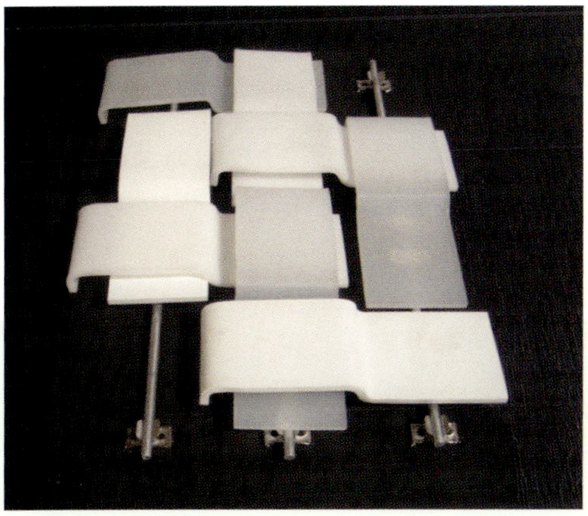

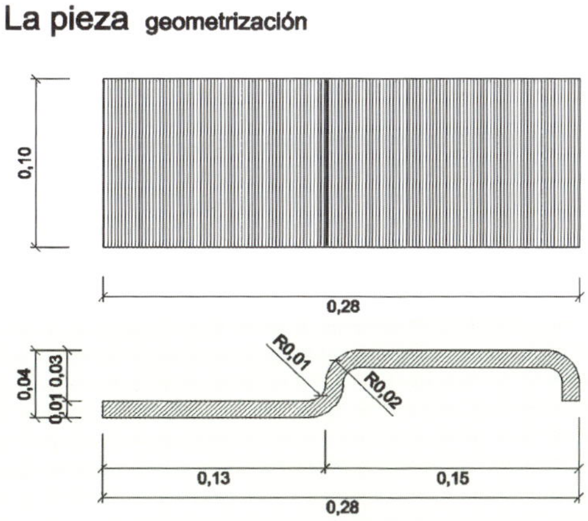

Onda (Wave)
Jorge Salinas

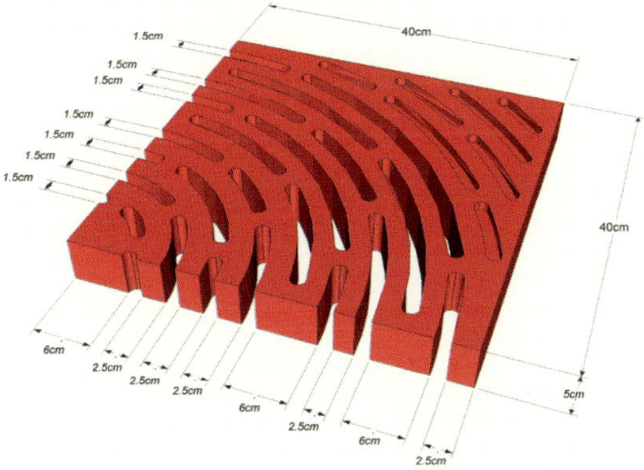

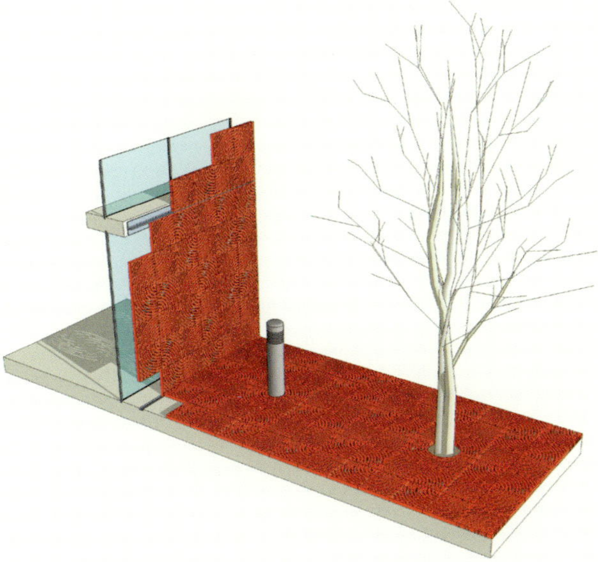

Cubilote (Cupola)
Maria Roca-Ribes

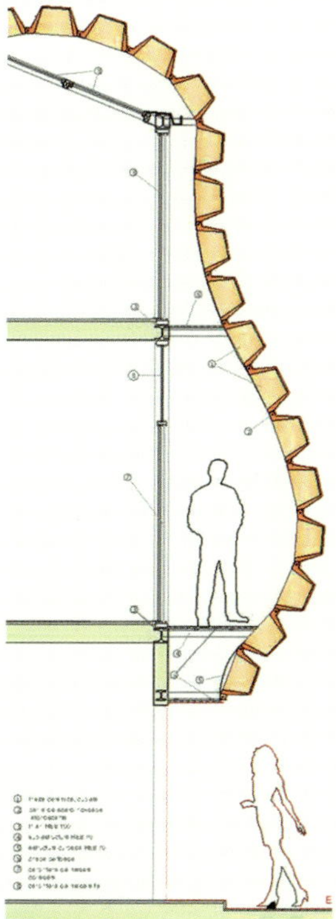

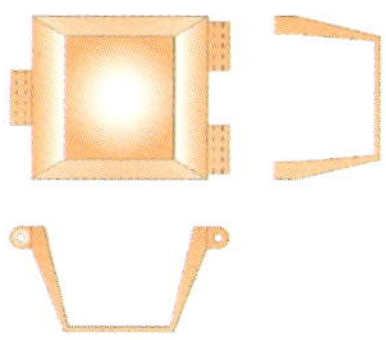

Also in the cobblestone format are the versatile Empty Brick by Xavier Martinell or the recycled Cantoline by Mario Blanco. Seeing as these contributions continue on from and offer a resolution for some projects from previous years, they could be presented as the third part of a trilogy – the trilogy of ceramic invention.

H2Orizón
Ester Laso

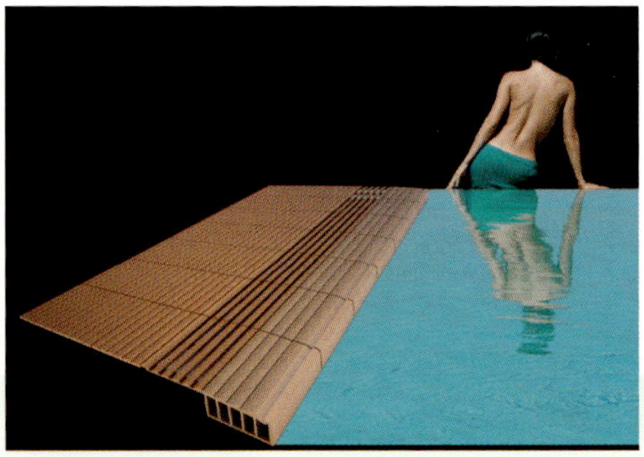

Empty Brick
Xavier Martinell

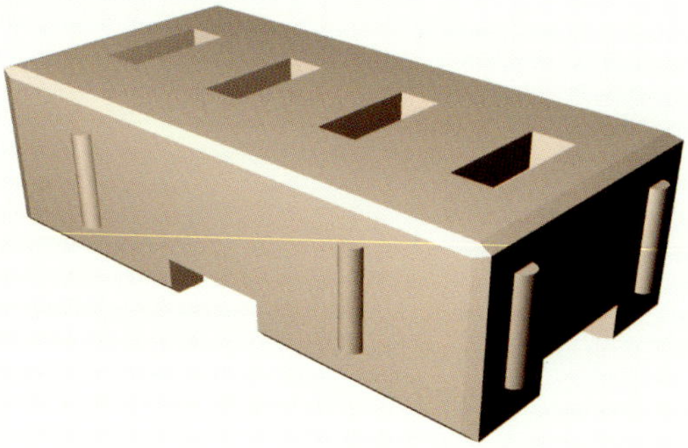

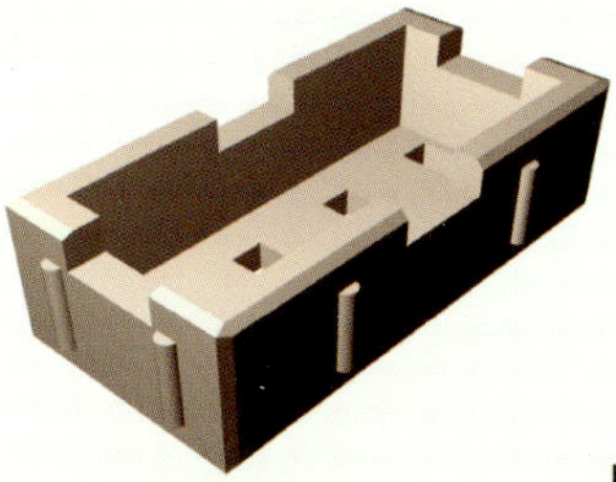

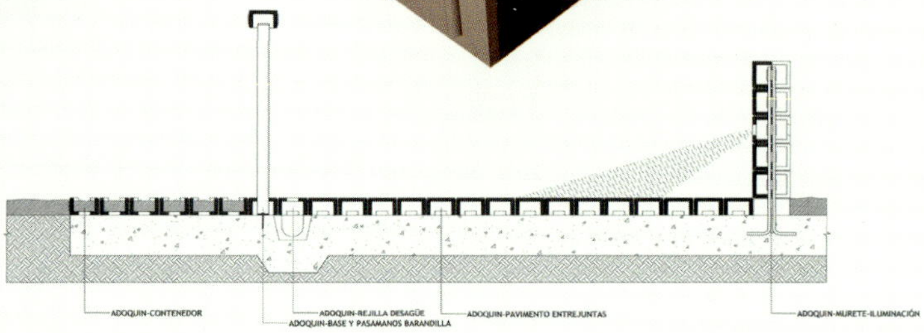

Cantoline
Mario Blanco

Eduardo de Miguel

Ceramic, materiality and identity

The Ceramic Studies Department at the Valencia School of Architecture

The new Ceramic Studies Department at the Polytechnic University of Valencia's School of Architecture, Sponsored by the Spanish Ceramic Tile Manufacturers' Association (ASCER), was set up with the purpose of disseminating architectural use of ceramics to the university community and to promote use of this material by researching new and original ways of working with it.
In order to do this, the Architectural Project was chosen on the basis that it is a creative process that allows intentions to be transformed into reality, achieving the complete transition of a design solution in response to a need into a transcendental masterpiece. We find ourselves in an unusual situation in which one specific material must be inevitably used in the setting out of the project. This material has an atavistic origin and specialisation with it has come as the result of an age-long process in which a good way of using it has always been found. At present, this evolution is not only far from running out of steam, but we actually find ourselves in the face of a new technological and creative renaissance.

Lorenzo de Navascués, Alicia Redondo, Alba Revert, Santiago Vicente

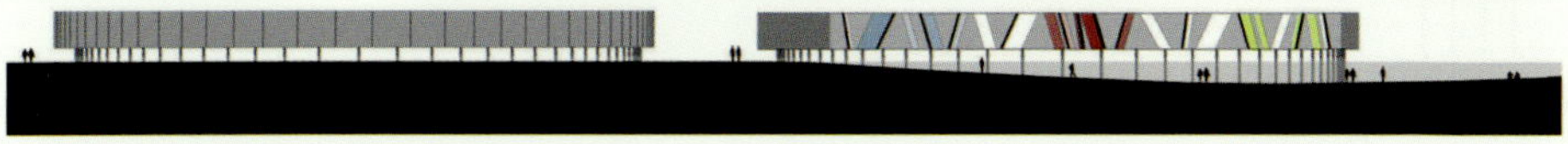

Miguel Bibiloni, Pau Boluda, Martín Méndez, Elvira Nebot

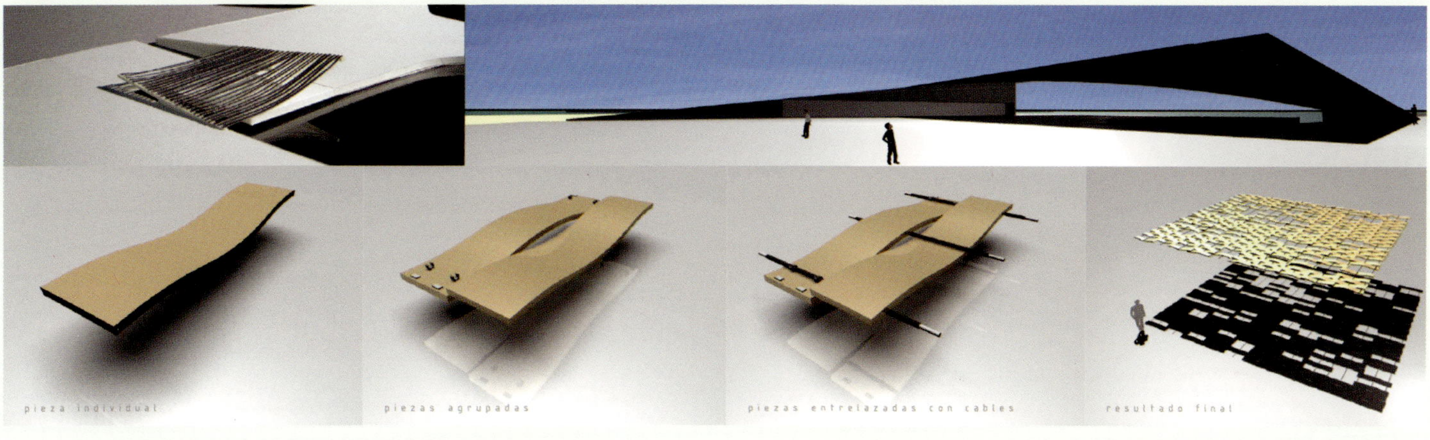

Diego de Ramón, Elena Pérez Borrel, Paula Quílez

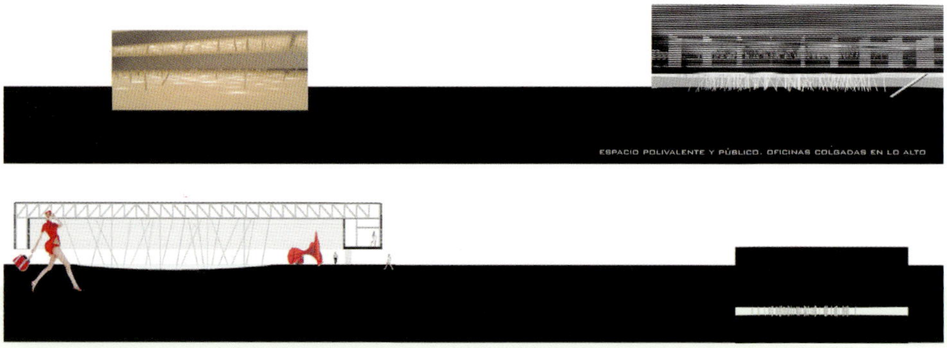

Estefanía Carreño, Ana Abellán, Pedro Gómez García

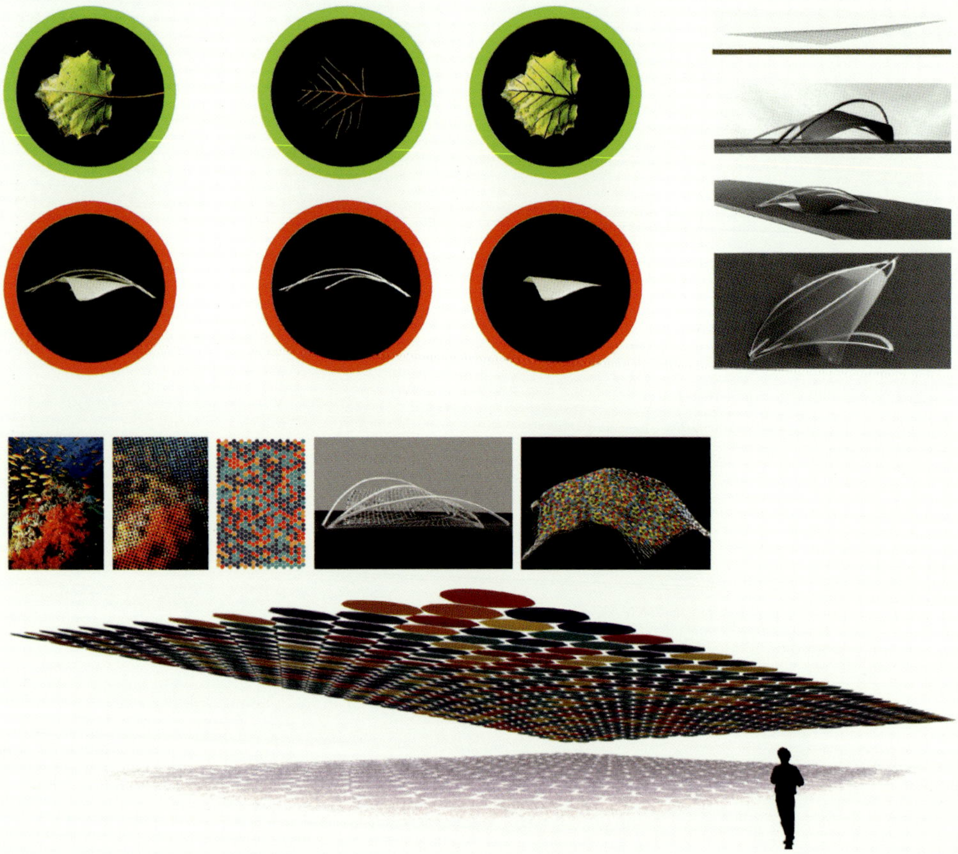

Logically, the first step to working with ceramic is to understand the nature of the material; to know where it comes from, what properties and limitations it has, and how it must be used. Therefore, it is fundamental to analyse the existing relations between the building process and production systems, sustainability as an unavoidable parameter in the definition of the project, and the rationalisation of the design and the necessary resources for its execution. It is more and more common to create personalised designs, given that some manufacturers offer the possibility of making them up. However, this is something only reserved for really special designs. Consequently, even as we are still before one of the industries that is most open to digesting innovative designs, taking on a new design in each project is totally unthinkable today.
From this perspective, it seems more interesting to research products from catalogues to see if, with slight variations, initially unforeseen architectural problems can be resolved, and to find the solution with our thoughts turned towards ceramic materials. A strategy for finding new applications to those already existing would mean to give greater protagonism to conceptual questions instead of exclusively formal ones, i.e. it would mean giving priority to project decisions – what – over design ones – how – even though both are essential, but in my opinion, the latter must be understood as consequences of the former. In order to reflect on the fundamental aspects that must be covered in an architectural project and the links that may be created with ceramic materials, it is appropriate that principles defined by Gottfried Semper in his book, The Four Elements of Architecture, should be used. They are: the hearth, the wall, the mound and the roof.

Mario Vega, Carlos Jiménez, Adela Ruiz

The element chosen to bring about all of these intentions was the roof. The aim of the project was to cover an area of 60 x 60 m, allowing different activities to take place beneath this roof. It was conceived of as an open, covered space without significant functional pre-requisites. It was not necessary to enclose the sides, nor was it important that rain should pass through it.
There are several arguments to justify the choice of these parameters. The fact that it was not important if rain passed through was purposely to escape from the idea of the roof as an element of shelter. By excluding this typology, it is not common to find designs made in ceramic; this fact allows research on what shape should be used or what variations must be made to existing materials to respond appropriately to the problem of covering a space. In other words, this proposition required rethinking and once again asking the question – what does if mean to cover a space? This approach enabled the project strategies, the real purpose of the design, to be broadened as much as possible and allowed two important issues to be avoided: the preconceptions many architects have about these materials, seeing as they had to be used in an unexpected way, and that the project would be turned exclusively into a construction problem.

Clara Arnau, Ligia F. Martins, Carmen García Jiménez, Belén Montalbán

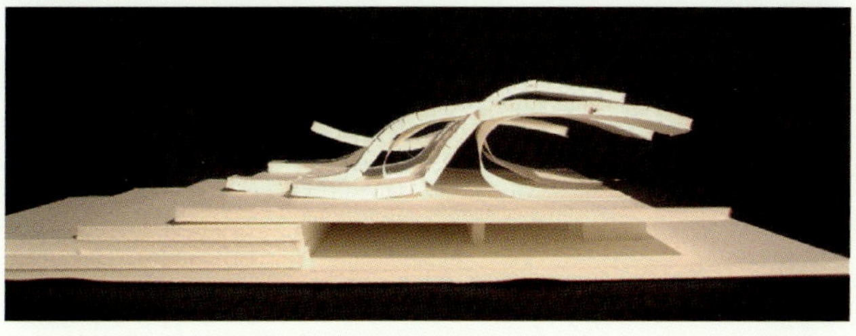

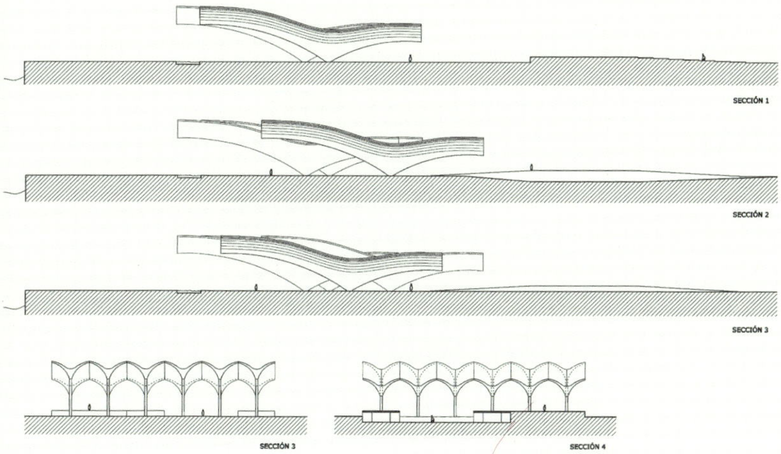

Alonso Balaguer and
Arquitectos Asociados
Lluís Alonso, Sergi Balaguer

International Center for Advanced Medicine / Wellness O_2

Barcelona, Spain

At the International Centre for Advanced Medicine (CIMA) and the Wellness O_2 complex, the concept of a sports club goes one step further to enter the world of health and well-being and to become not only a medical and sports reference, but also an architectural landmark for the district and the city. This new concept is reflected by its name – Wellness Centre – and this has also been applied to a new notion of sport-related health. The floor plan of these centres gives maximum flexibility for multi-purpose use. Different levels are dedicated to medical consultation rooms, hospital inpatient rooms, operating theatres and complementary areas. The ground floor is the setting for the public areas such as the spacious foyer and bar-restaurant with an independent street entrance.

The medical centre was conceived as a large, shiny and clear glass box - a jewel set into the ground. Its aspect changes throughout the day owing to the ever-changing reflected sunlight. The external, transparent image speaks of the clean, aseptic interior, while users are able to enjoy breathtaking views of the city and the adjacent park.

The interior layout also differs greatly from the image of a conventional clinic and creates a new conceptual line that achieves a positive effect in the users of the building and the services on offer to them.

The number of different materials used have been kept to minimum in a bid to create an all-encompassing vision of space.

The colour and texture of the ceramic tiling unifies all of the special uses. It is also present in spa baths, saunas, steam baths and so on.

The ceilings are covered in tiles made of pressed shavings painted the same colour as the ceramic tiles.

All of the spaces are laid out as free-standing boxes, creating interesting flows around them.

The main pool is lined with stainless steel die stamped plates with a spill-over side and creates a visual impact by causing the viewer to think the water goes over into the adjoining park, an effect sought by the architects.

Graphic design produced by the same team of architects enhance the concept of reducing messages to a minimum by incorporating them into the architectural design.

The desired quality and warmth are likewise expressed in the changing rooms where soft lighting accompanies the laminated wood furniture that was specially-designed for the sports chain.

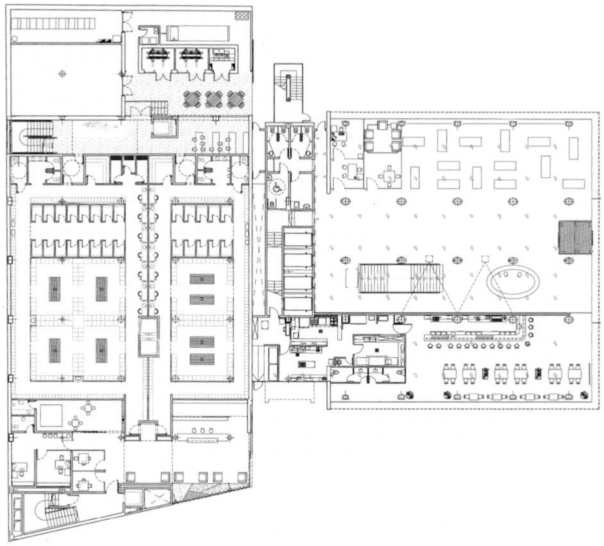

GROUND FLOOR

FIRST FLOOR

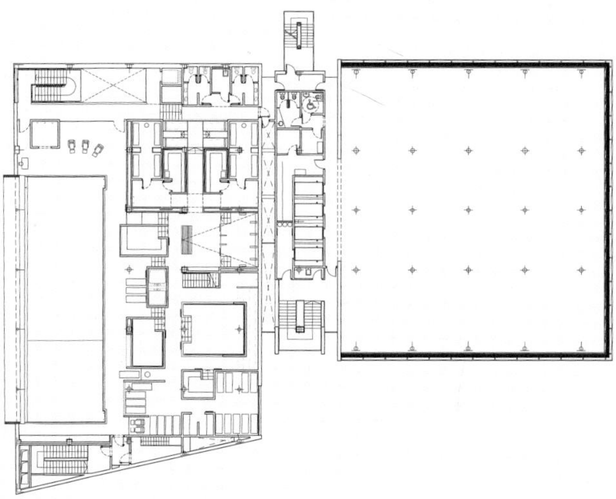

FOURTH FLOOR

ARCHITECTS: Alonso Balaguer y Arquitectos Asociados. Lluís Alonso, Sergi Balaguer / **LOCATION:** c. Manel Girona – c. Eduardo Conte, Barcelona / **CHRONOLOGY:** 1999 – 2002 / **CLIENT:** Metrópolis Sports Club, s.a. / **ENGINEERING:** INDUS, s.a. / **INSTALLATIONS:** Construcciones Rubau, s.a. / **BUILDER:** Construcciones Rubau, s.a. / **QUANTITY SURVEYORS:** Enric Llucià y Agustí Villa Isorna / **BUILT AREA:** 22.000 m² / **TOTAL COST:** 10.000.000 € / **PHOTOGRAPHER:** Jose María Molinos

CERAMIC: Ceramic tile in diferent sizes

EMBT Arquitectes
Enric Miralles, Benedetta Tagliabue

Park and Civic Center

Mollet del Vallès, Spain

The park and civic centre, important places for locals to gather, are located in an outlying suburban area.
From the outset, an existing Romanesque building served as the model inspiring the new centre. Although originally meant to be a copy of the older building, this gradually began to change and settled into its setting, only keeping the long stairway and access ramps.
Pictures cover its walls.
The same walls that define different areas of the park are transformed into graffiti, repeating the name of the place and sequences of letters, which, piled one over the other, throw shadows onto the ground.
The many shadows from the building define space on the ground and small changes in surface contours make possible the incorporation of walkways, play areas, areas for skating and skateboarding and petanque, etc.
Vegetation is present in the form of compact garden areas and is another element that aids in the layout of the different spaces.
Water appears and disappears during the daylight hours, occupying surfaces designed with blue glazed ceramic tiles.
These are large organically-shaped basins in with bright tiling that accentuates the protagonism given to abstract and artificial composition in the park's lay out. When the fountains are not in use and the surface is dry, their design and the warm quality of the material allow them to be used as benches.

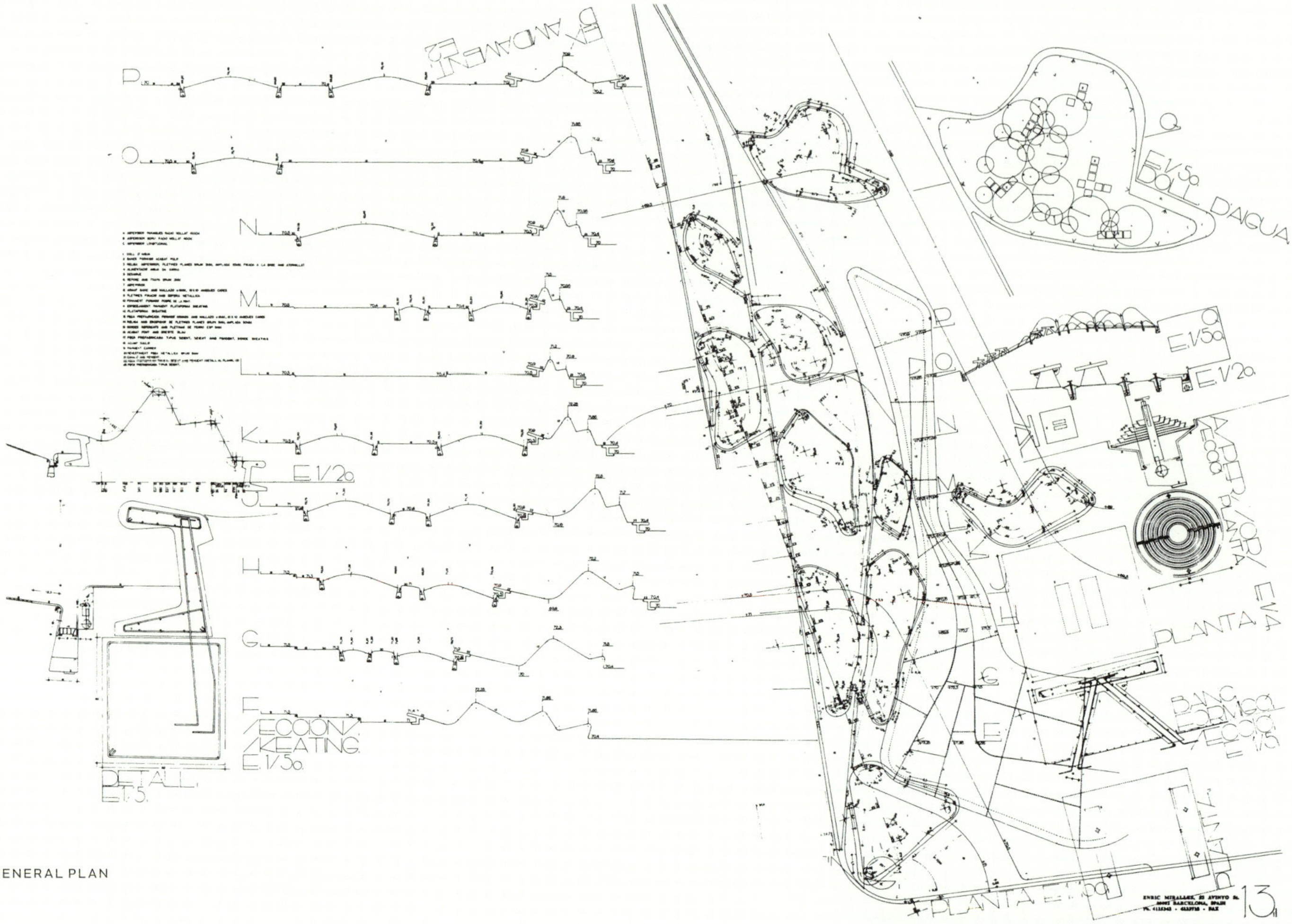

GENERAL PLAN

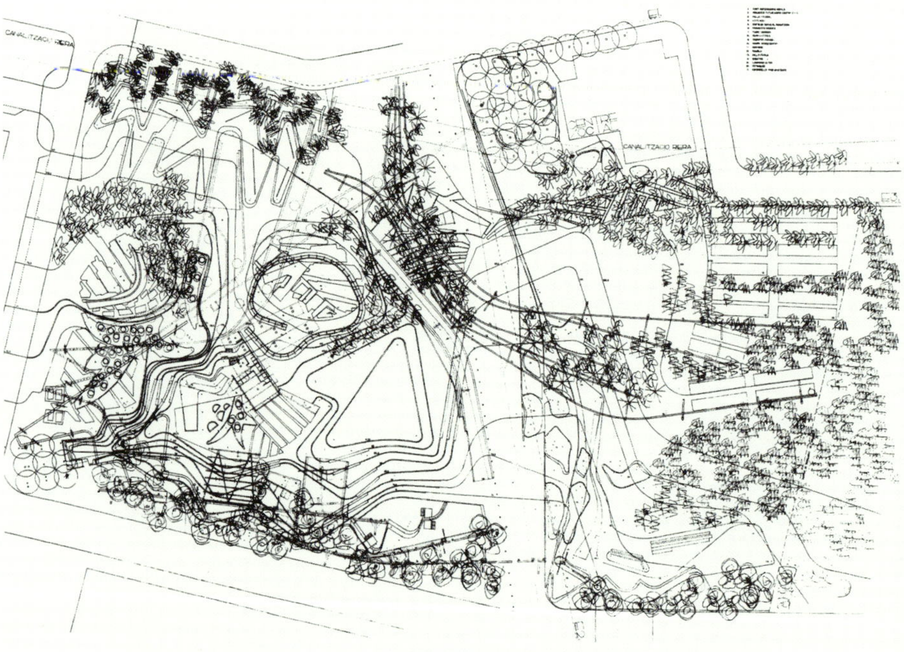

DETAILS CERAMIC FOUNTAINS

ARCHITECTS: EMBT Arquitectes. Enric Miralles, Benedetta Tagliabue / **LOCATION:** Plana Lledó-Mollet del Vallés, Barcelona province / **PROJECT:** 1992 – 1995 / **CONSTRUCTION:** 1995 – 2001 / **CLIENT:** Mollet del Vallés City Council / **PRELIMINARY PLAN:** Ricardo Flores, Josep Cargol / **COLLABORATORS:** Joan Callís, Josep Miàs, Ricardo Flores, Josep Cargol, Jordi Artigues, Mary Rose Greene, Lucía De Colle, Nicolás Alvarez, Victoria Garriga, Sibyl Maurer, Germán Zambran, Hubertus Poppinghaus, Umberto Viotto / **MODEL:** Pia Wortham, Miquel Lluch, Annie Marcela Henao Sanabria / **PROJECT MANAGEMENT:** G.3, Jordi Altés, Dep. Servei Tècnic Ajuntement de Mollet del Vallés / **ENGINEERING:** OVE ARUP & Partners, Miguel Barreras, CEA, J.C. Adell, Ricard Pi, Ricardo Serra / **INSTALLATIONS:** PGI, Josep Juliol and Josep Masachs / **BUILDER:** COPCISA Laia Codina, ALM-4, BONAL / **BUILT AREA:** 20.000 m²

CERAMIC:
Fountains lined with specially-designed blue glazed tiles

Guillermo Vázquez Consuegra

Castilla-La Mancha Regional Archives and Bibliographical Repository

Toledo, Spain

Winner at 2006 Tile of Spain Architecture and Interior Design Awards

The Castilla-La Mancha Archives can be understood as a huge chest, a strongbox that fiercely guards the cultural heritage of the Autonomous Region. This image was precisely what was envisaged for the building: a compact, closed, free-standing and inward-looking structure for the safe-keeping of the treasures deposited within its walls. It is a building that does not show off the diversity of its interior spaces; one that maintains a discreet relationship with its setting while making its solid and efficient presence felt. Underlying this figurative image that has been taken on by the design is the idea of stratification, of layer placed over layer, likening it to a pile of documents. The building is closed and heavy but perceptively lightened by means of its strata composition. The design rejects the idea of defining the Archives as merely a repository of documents or as a storage facility full of shelves. On the contrary, it proposes that the interior space should give meaning to the building. It is an ample and fluid space, essentially defined by a great void that passes through all of the building's levels and gives natural light an important role, an atypical aspect in buildings of this nature.

The building is composed of five storeys above ground and one below ground level. Common, public areas plus those housing technical processes, management and administration of the Archives are located on the ground and mezzanine floors, while services to the general public are housed on the lower ground floor, opening onto a garden dug at that level. Planted with aromatic shrubs, it is an enormous carpet of French and common lavender. Trees of a larger size are found elsewhere on the lot.
The Repositories are located on the four upper levels. The standard floor plan is a grid made up of 42 modules laid out in such a way that the central void passes through the whole building, separating the Regional Archives from the Bibliographical Repository, and manages the relationship between these areas and the complementary, common areas of the building through two vertical communication cores. The façade of the building is clad in tiles that express local tradition. Two types of tile were exclusively designed, both of a large size but one being flat and rectangular and the other in the shape of a T. This combination enabled the different requirements of the various façades to be resolved. This cladding is nothing more than a delicate wrapping of the building in horizontal tiles to unify the façades and the roof with one material. At the same time, these walls change to express the mood created by the different conditions created by weather and sunlight.

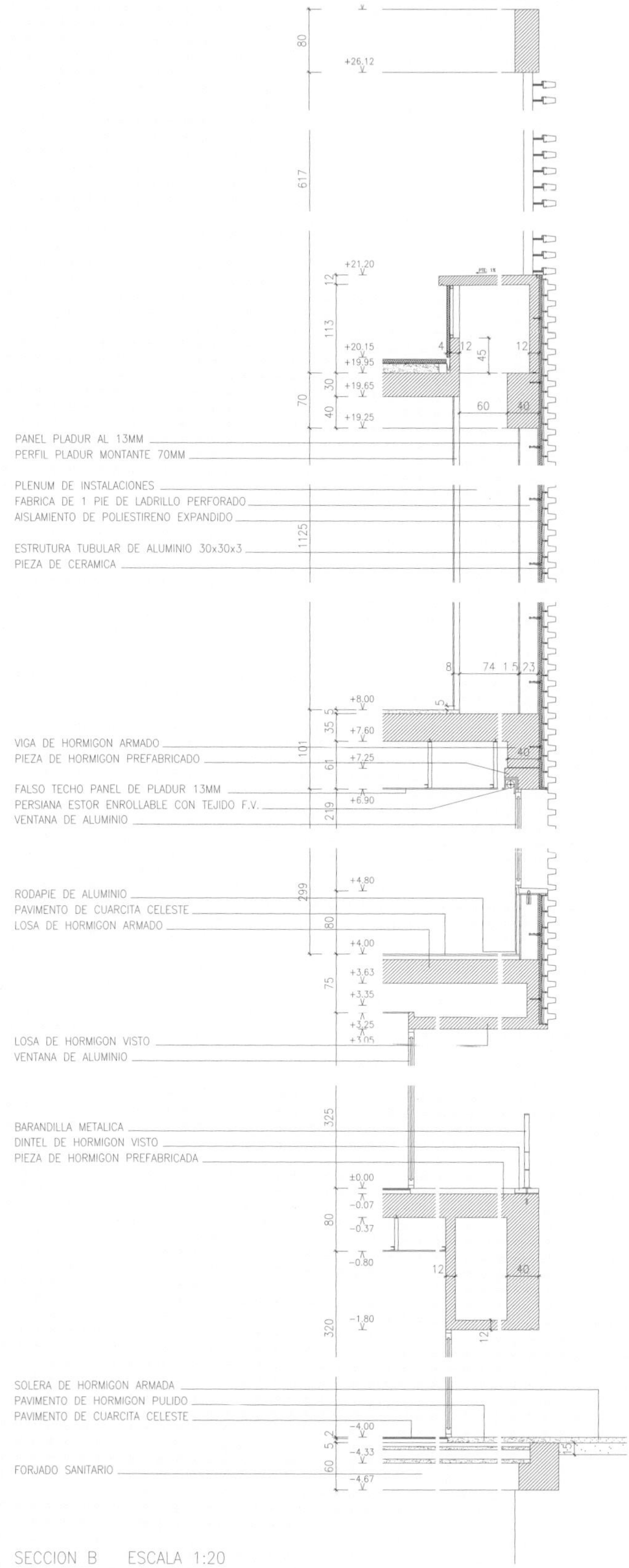

CROSS-SECTION FAÇADE. CONSTRUCTION DETAIL

abida en el Archivo de Castilla-La Mancha.
Un edificio que no quiere ser guardián de memorias pretéritas sino el testimonio vivo y presente de

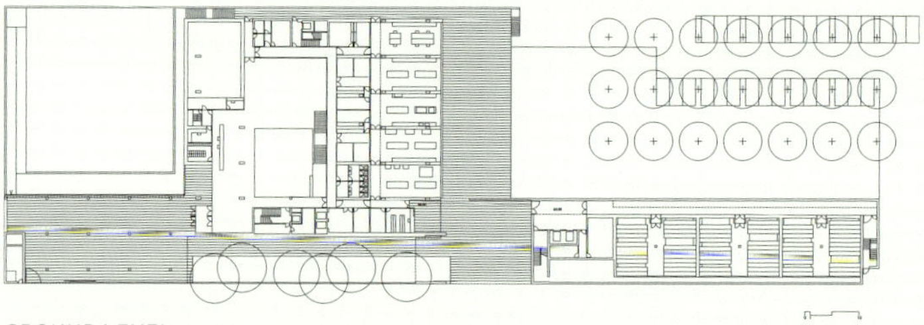

GROUND LEVEL

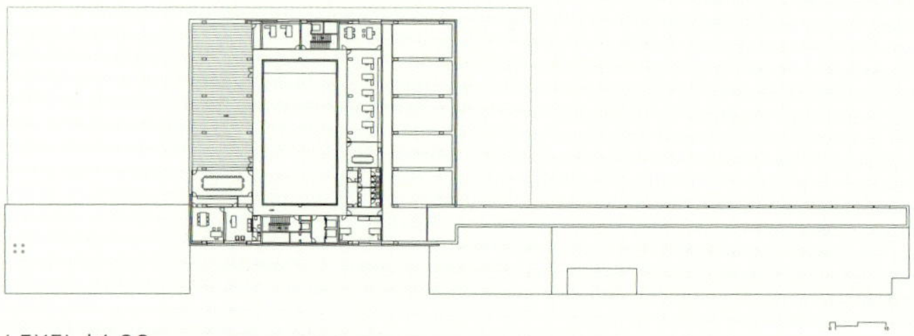

LEVEL +4.00

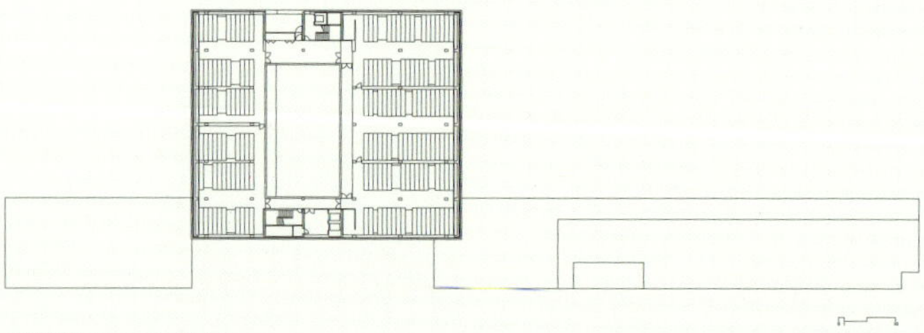

TOWER FLOOR PLAN

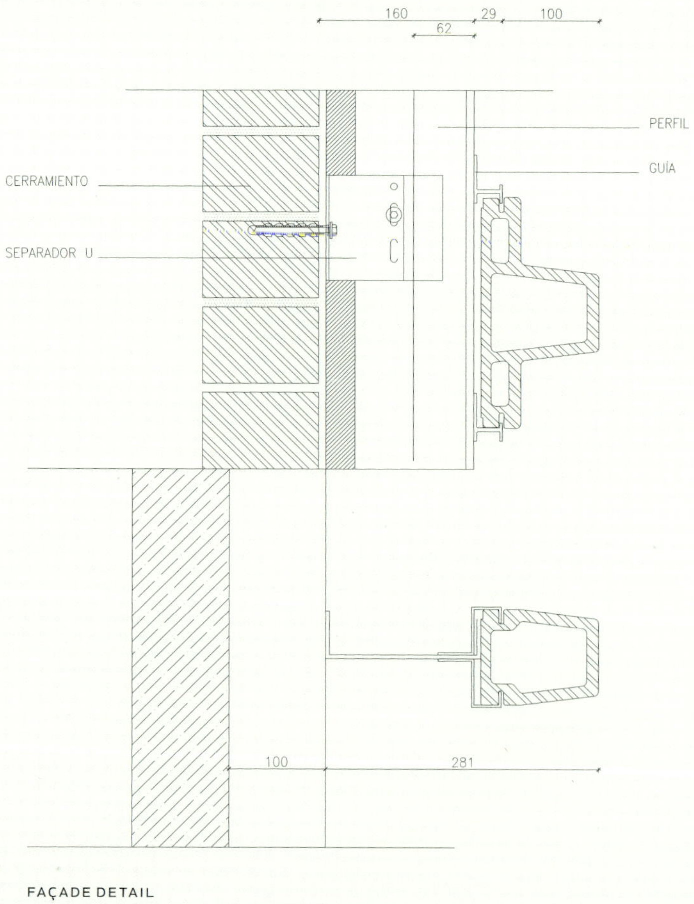

FAÇADE DETAIL

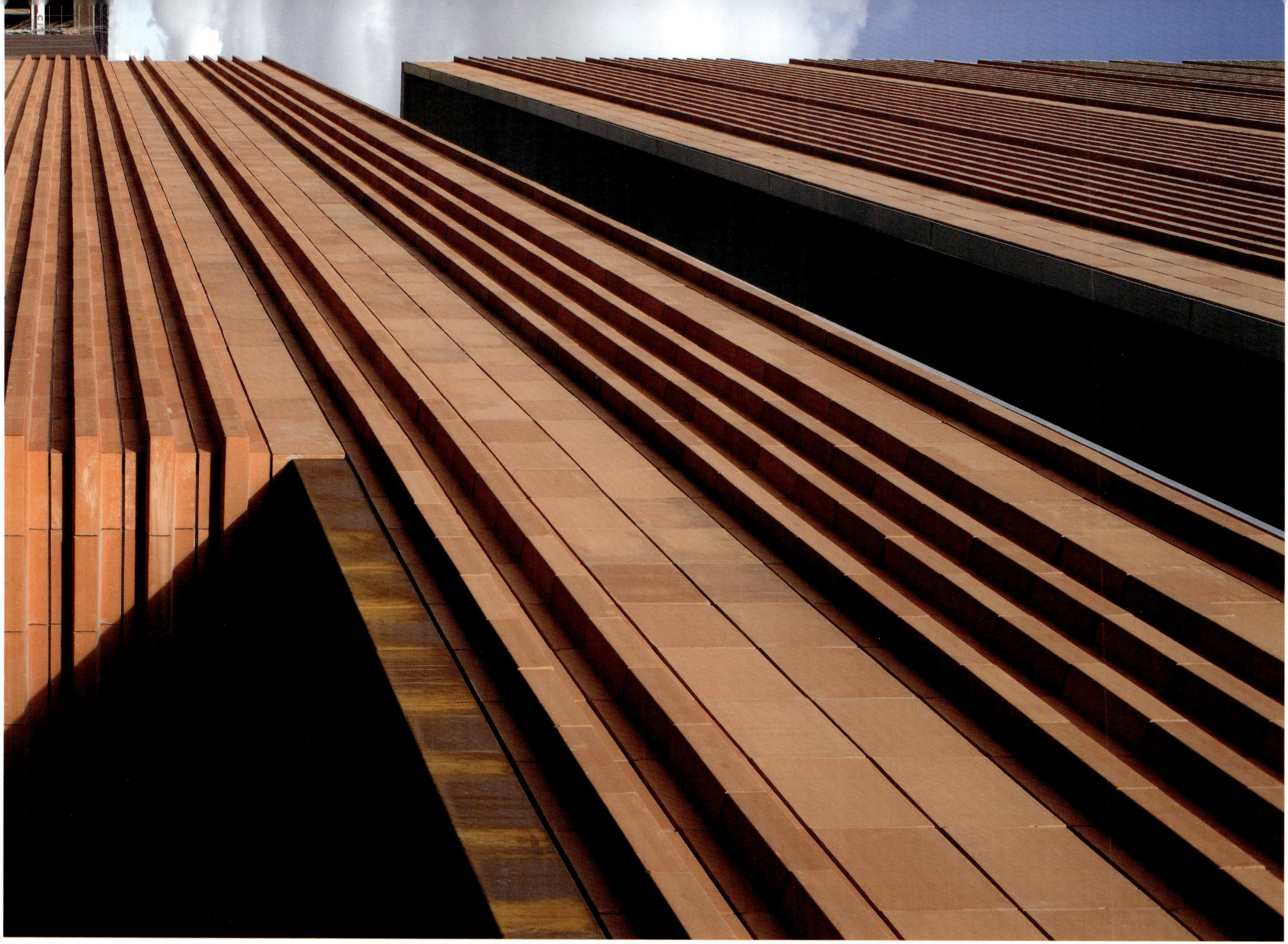

ARCHITECT: Guillermo Vázquez Consuegra / **LOCATION:** Polígono de Toledo / **COMPETITION:** 1997 / **PROJECT:** 1998 / **CONSTRUCTION:** 2000 – 2005 / **COLLABORATORS:** Marcos Vázquez Consuegra (ATCC), Frank Mazzarella, Juan Fernández-Andrino, Romek Kruszewski, Miguel Chaves / **CLIENT:** Junta de Comunidades Castilla-La Mancha / **QUANTITY SURVEYOR:** Juan Valverde, Eusebio Sánchez / **FRAME:** NB-35, s.l., J. Jiménez Cañas / **SYSTEMS:** INSUR-JG, s.l. / **MODEL:** Frank Mazzarella / **FAÇADE CERAMIC TILES:** Vázquez Consuegra / **BUILDER:** Necso Entrecanales Cubiertas, s.a. / **BUILT AREA:** 12.544,64 m² / **TOTAL COST:** 12.652.976,08 € / **PHOTOGRAPHER:** Duccio Malagamba

CERAMIC: External cladding with large horizontal tiles, one type flat and rectangular and the other T-shaped, attached with metal staples to tubular aluminium frame, the combination of which allows the different façade requirements to be satisfied

Teresa Sapey

100% Cerámica Espacio ASCER Casa Decor Madrid '06

Madrid, Spain

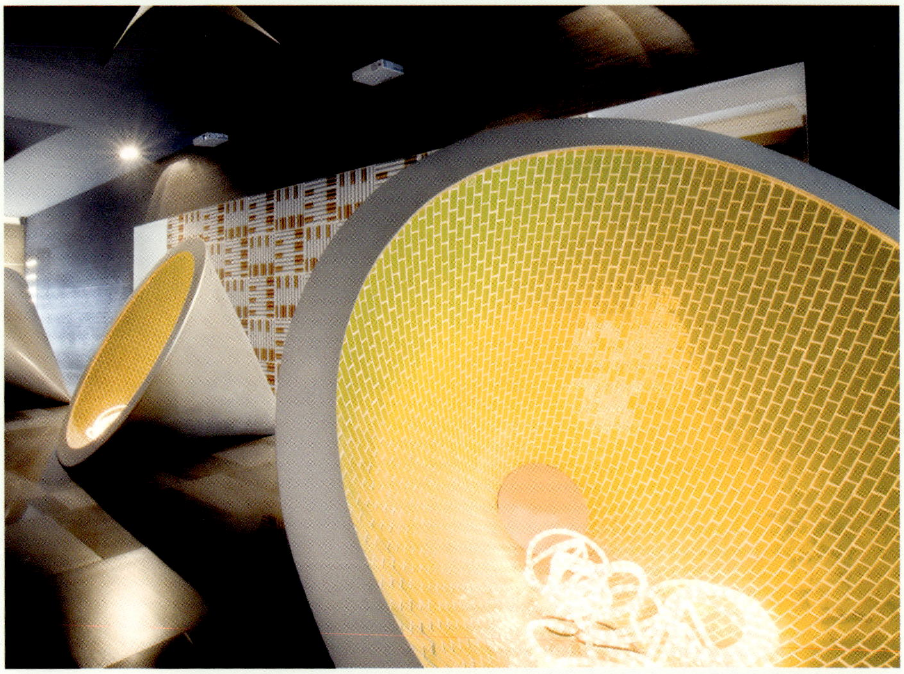

This project for the ASCER space of Casa Decor was based on the idea of being able to display a great deal of new covering products in one place. A concept was set out that would act as a main structure on which an image would be created that would not be fragmented.
Starting from the representation of smoke — that can be understood as metaphor for knowledge — the idea was to incorporate new technologies (video projections, printed ceramics, among others), and likewise new ways to understand furniture, to extend the concept of the exhibition space.
To achieve this, cone-shaped seats covered with small tiles were designed to function as "islands" on which to sit down to rest (thus disconnecting ceramics from their traditional use as floor coverings). While sitting, visitors could observe a dual-channel projection of smoke (alluding to classic cinema, especially to the film noir genre).
On another wall, however, the smoke was interpreted geometrically, creating a drawing from horizontal stripes that gave way to luminous tubes (fluorescent). A reticular graphic structure was created from a photograph of a cigarette that covered the wall in front of the video projection.
The result was a space that differs from traditional product publicity stands. Instead, ceramic material was used to situate the public in a dream-like and comfortable setting.

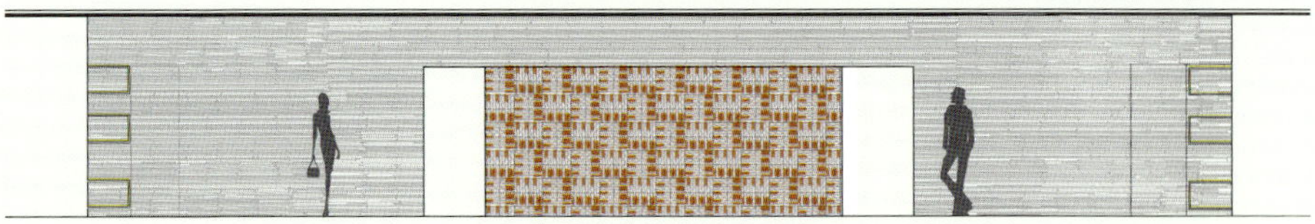

LONGITUDINAL SECTION

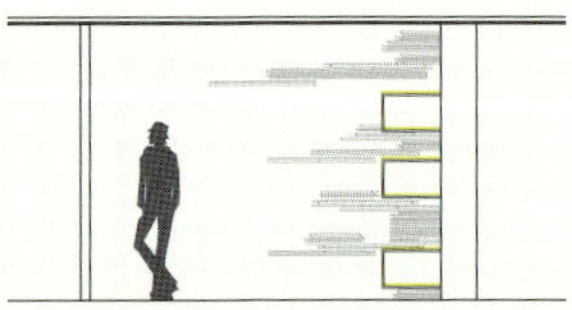

TRANSVERSE SECTION

ARCHITECT: Teresa Sapey / **LOCATION:** Salustiano Olozaga 11, Madrid / **CHRONOLOGY:** 2006 / **COLLABORATORS:** Ángela Sanz, Marta Melendo, Daniela Calvo, Raquel Rojas, Alex Gutwil / **CLIENT:** ASCER / **BUILDER:** Litecon

CERAMIC: Porcelain stoneware, mosaic and glazed tiles in different sizes

Daniel Buren

"Les Cabanyes de ceràmica i espill" Installation in the Espai d'Art Contemporani de Castelló

Castellón, Spain

Between the constructive simplicity of the first *cabanes éclatées* (exploded cabins) created by Daniel Buren at the beginning of the eighties and the complexity that is demonstrated in his most recent ones, these architectural structures created by the artist throughout last the twenty-five years continue to form efficient spatial perception devices for the visitor. They do so equally in relation to the architecture with which they fusion and with the geometry that shapes their planes, angles, openings and projections.

For *Les Cabanyes de ceràmica i espill* – a project specially designed by Daniel Buren for the Castelló EACC – the exhibition space comprises six large cabins of 5.65 x 5.65 metres placed on either side of the central mezzanine of the building.

The individual position of each cabin varies, some being arranged parallel to this central structure, while others "attack it" from different angles. Through these cabins, visitors are presented with a new route (both on the ground and first floor) and are led through a series of doors in each one, which are physically projected on the outside of each cubicle. Thus, a new and powerful intermediate ("between") space is generated that completes the common interior/exterior dialectic of any building.

Unlike other similar works that emphasise the outside of the cabins, on this occasion each cabin has been completely painted white. No other

architectural concessions have been made other than their own cubic structures. Nevertheless, the interior of each cabin is totally covered with ceramic material. In turn, each of the four walls has a special design consisting of a variety of coloured planes and shapes, not missing the trademark 8.7 cm wide alternating stripes, which the artist has incorporated in his work since the sixties.
Moreover, two of these cabins have two interior walls covered with mirrors that double the interior space, accentuating the idea of continuum. A seventh architectural structure forms a porch of the double route. It begins in the outside of the museum and penetrates into the reception area, leaving again at a right angle to the exterior. This gives the idea of a cabin that is open to the city space and definitively establishes a link between the museum and the public space.

CUBES FACE A-C

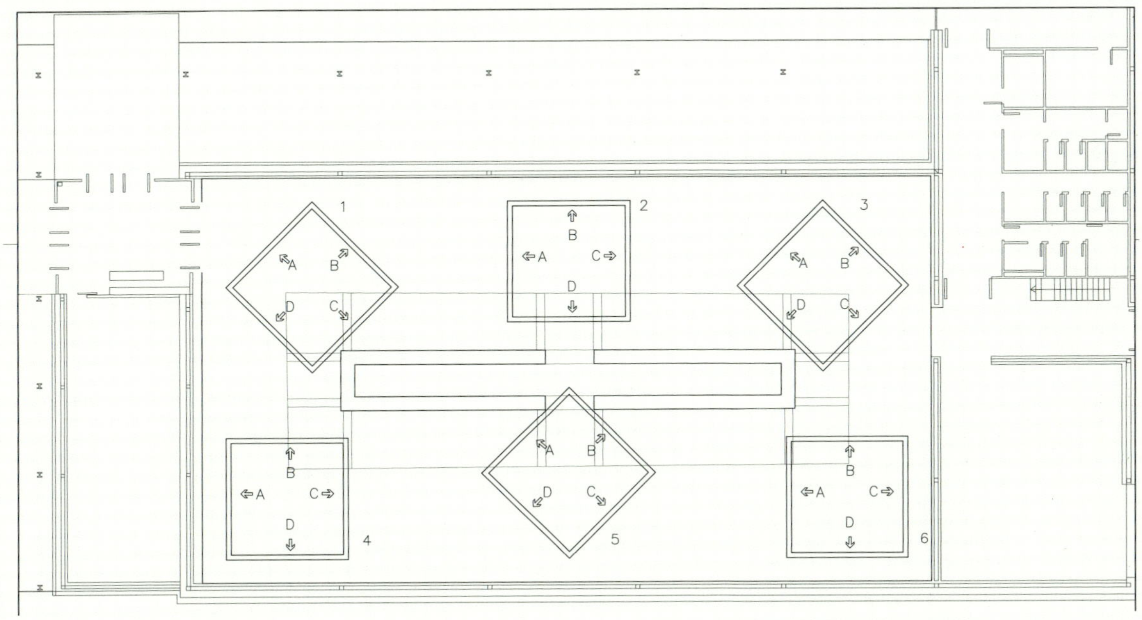

GENERAL PLAN

ARTIST : Daniel Buren / **LOCATION :** Espai d'Art Contemporani de Castelló / **CHRONOLOGY :** 26 de abril – 24 de Septiembre 2006 / **CONSTRUCTION :** COARCE (tile cutting and meshing), Ernesto Boira (Alicatado), JM Trans (Montaje) / **COLLABORATORS :** ALICER, Instituto de Promoción Cerámica (institute for the promotion of ceramics), Association Française d'Action Artistique (french association for artistic action), Institut Français Valencia (the french institute of valencia)

CERAMIC : Single-colour red body tiles, shiny and matte (approx. 20.000 tiles)

Javier Neila, César Bedoya,
Alfonso García-Santos,
Estefanía Caamaño,
Miguel-Ángel Egido,
Javier Jiménez,
Luis Magdalena
and degree and
postgraduate students
from the Polytechnic
University of Madrid

Magic Box "Decathlon" prototype for an energy-efficient sustainable home

Washington D.C., USA

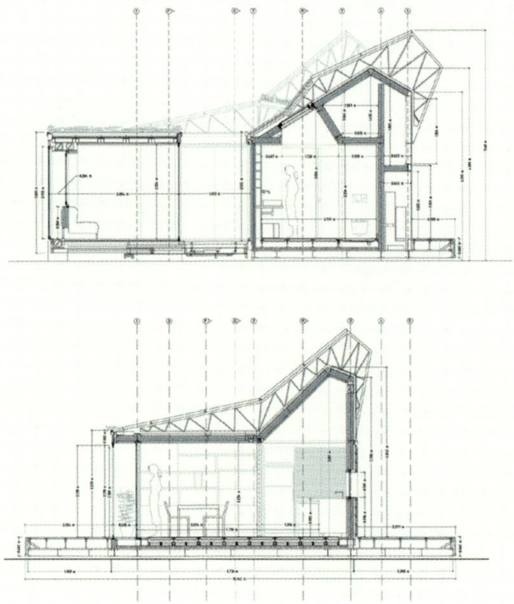

TRANSVERSAL SECTIONS

The polytechnic university of Madrid was the only European representative in the second edition of the international competition "solar decathlon" run by the us department of energy. Its final phase was held between 29th september and 19th october 2005 in the city of Washington D.C. The event gathered together 18 teams representing the American and European universities.
It consisted in designing and constructing a 21st century detached house that was powered only by solar energy.
The name "decathlon" reflected the nature of the competition that comprised ten tests in which the participants, called decathletes, had to present their entries.In the final prototype demonstration phase that was held in the national mall of Washington, the viability to carry out everyday tasks was tested before professionals from the fields of architecture and engineering, the media and the general public.

The ten competition tests were based upon three basic principles:

I. Supplying the energy needed to carry out day-to-day tasks (feeding, cleaning, leisure, work, transport, among others) with an acceptable level of comfort and using only the solar power collected by the house during the seven competition days.
II. Demonstrating to society, using a practical approach, the existence of architectural design principles that use solar technologies and their aesthetic and energetic benefits.
III. Stimulating research and development in renewable energies and energy efficiency, particularly in the construction sector.

Quality of life was the inspiration behind the Spanish entry that was christened Magic Box. Thus, special attention was paid to aspects such as air quality, thermal comfort, humidity and the correct distribution of the interior temperature. This theme was highly relevant as the competition was based on minimising energy needs through the use of bioclimatic design principles from Spanish traditional popular architecture, optimised by integrating modern technologies. Ceramics were used in keeping with the goal of reaching the highest possible levels of environmental sustainability. They played an important role in the construction of the ventilated façade and in the interior stoneware floor, providing the final visual finish of the box.

ELEVATIONS

ARCHITECTS: Javier Neila y César Bedoya (architecture), Alfonso García-Santos (construction), Estefanía Caamaño y Miguel-Ángel Egido (photovoltaics), Javier Jiménez y Luis Magdalena (domotics), Degree and Postgraduate Students of the Polytechnic University of Madrid / **LOCATION:** Temporarily Located In The National Mall of Washington (September – October 2005). Currently model 1 is in the Campus of the Polytechnic University of Madrid and model 2 is in beijing, in the Future House site / **PROJECT:** 2004 – 2005 / **CONSTRUCTION:** Los distintos modelos y en distintos emplazamientos entre 2005 y 2007 / **CLIENT:** built at the request of the us Department of Energy, but self-financed / **BUILDER:** Students of the Polytechnic University of Madrid: Alfonso García-Santos (construction), Estefanía Caamaño y Miguel-Ángel Egido (photovoltaics), Javier Jiménez y Luis Magdalena (domotics), Degree and Postgraduate Students of the Polytechnic University of Madrid

CERAMICS:
Ventilated façade of porcelain stoneware tiles and stoneware floor tile

Sara Cole

Floating House dwelling-refuge

Bristol, United Kingdom

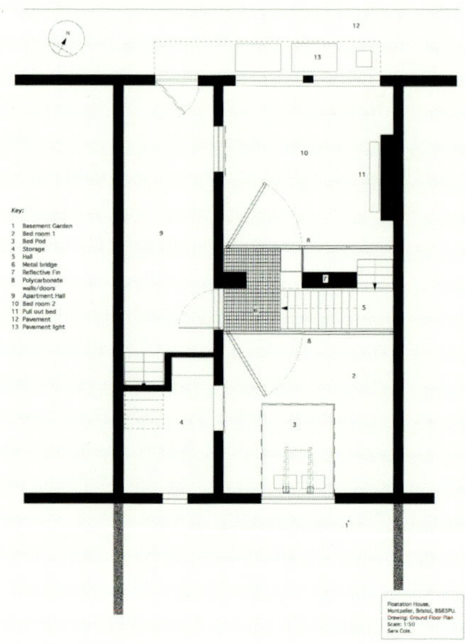

GROUND FLOOR

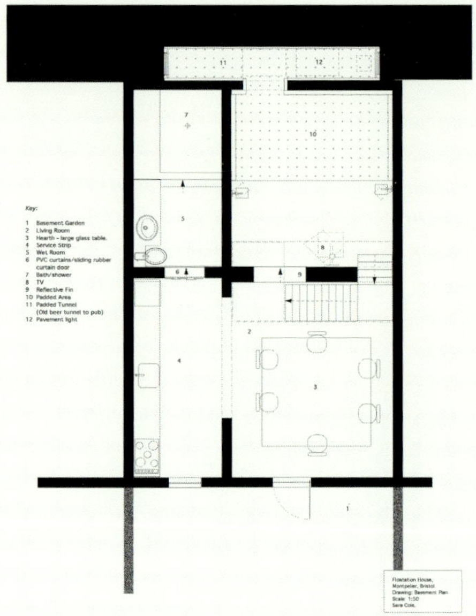

SEMI-BASEMENT

Floating house was designed as a tranquil retreat from the stress and noise of fast-moving city life It is a conversion of a former pub on the ground floor and semi-basement of a Georgian building. The design of the space brings a light atmosphere and it is saturated in blue tones resulting from the coloured PVC used as room dividers. The different areas of the dwelling are laid out around a white, tiled wall the height of the two floors that acts as a light-reflecting wall. White walls and homogeneous use of finishing materials enhances the timeless quality of the refuge.

CERAMIC : WHITE TILES 15 X 15 CM ON FLOOR, WALLS AND CEILINGS. BRIGHT WHITE TILES 20 X 40 CM ON DOUBLE HEIGHT WALL

ARCHITECT : SARA COLE / **LOCATION :** 38 BATH BUILDINGS, MONTPELIER, BRISTOL BS65PU. UNITED KINGDOM / **CHRONOLOGY :** 2001 – 2002 / **BUILT AREA :** 83 m^2 / **TOTAL COST :** 82.000 € / **PHOTOGRAPHER :** HUNTLEY HEDWORTH, RED COVER

XNF Arquitectes
Xavier Bustos,
Nicola Regusci, Ferran Grau

FAD Crypt

Barcelona, Spain

An unused crypt was found underneath the Exhibition Hall in the basement of the head office of Fostering Arts and Design (FAD). It was vaulted, with 39 tomb recesses and an annex. The project to develop this unique space consisted in creating a commemorative area to celebrate the FAD's centenary that would later be incorporated it into the Exhibition Hall. The determining factors of the project design were the small size of the crypt, the tomb recesses and the entrance – a staircase from the Exhibition Hall.

The project started at the crypt entrance where a glass floor panel was installed in the Exhibition Hall thus allowing the inside of the crypt to be seen. From here, the entry to the staircase accessing the crypt below is opened using a hydraulic piston.

The crypt was transformed by using a continuous covering of white Gresite tiles in a red grid, starting at the staircase and spreading across the floor, walls and vault of the central area. This gives visitors the notion of being in the heart of the FAD. All of the tomb recesses, together with an existing pit and annexe, are incorporated into the central area and painted in red. These were all transformed into unique display areas. The transformation was completed by placing a layer of glass in front of each recess. This acts as a support for backlit images and protects any objects on display.

The lighting inside the niches, pit and annex redefines the image of the original tomb.

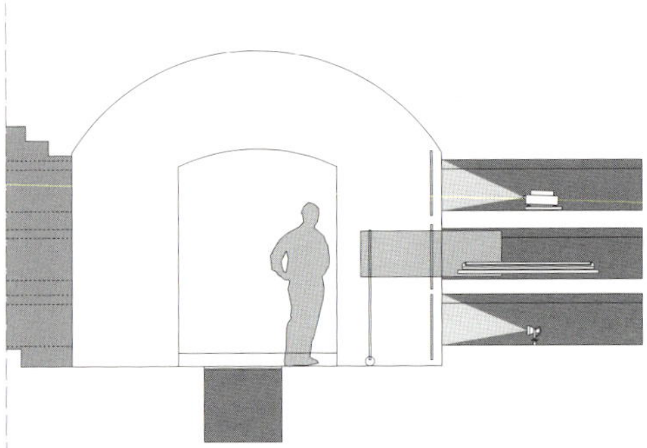

SECTION DETAIL

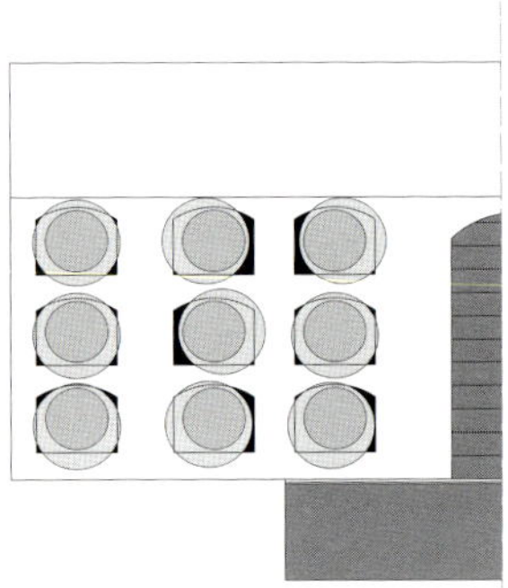

ELEVATION DETAIL

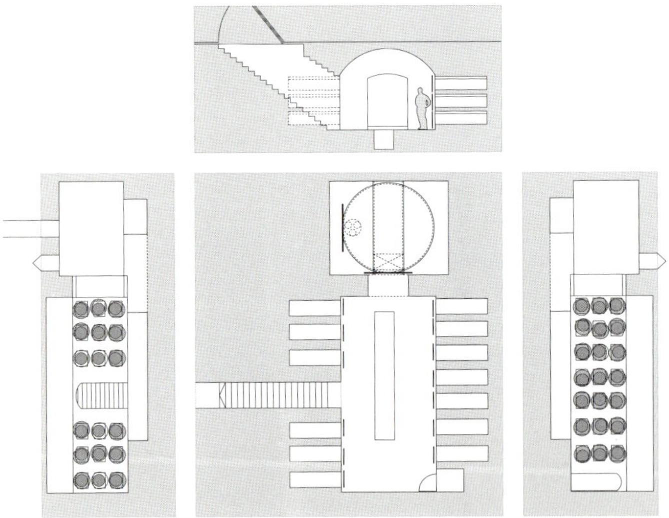

FLOOR AND SECTIONS

ARCHITECTS: XNF Arquitectes. Xavier Bustos, Nicola Regusci, Ferran Grau / **LOCATION:** Plaça del Àngels 5-6, Barcelona / **CRONOLOGY:** 2003 / **CLIENT:** FAD / **BUILDERS:** IGuzzini, Panasonic, Cricursa, Sistemas TDM, Arena, Pavindus, Technal, Sapic, TMP / **SURFACE AREA:** 60 m² / **PHOTOGRAPHERS:** Débora Tenenbaum, Rafael Vargas

CERAMICS: Continuous covering of white gresite tiles in a red grid, starting at the staircase and spreading across the floor, walls and vault of the central area

Patxi Mangado

Pabellón de España EXPO ZARAGOZA 2008

Zaragoza, Spain

The basic objective when designing the pavilion was to replicate a poplar or bamboo grove upon a water surface.
The results create a structure with enormous possibilities for a logical approach to energy and environmental commitment. These were both fundamental and emblematic for the Spain pavilion in the exposition. The architecture also makes the pavilion one of the most attractive physical and well-lit spaces that one can imagine.
The infrastructure organisation of the building works like a circuit of structural and energetic flows. Two great planes shape the building: one is the water plane from which the ceramic pillars arise, whereas the other is the roof — where the pillar reach up to. Between both there are a series of power circuits that easily and logically provide solutions for ventilation, air conditioning, etc. The vertical elements are made in a workshop. They have a metal core lined with equal ceramics tiles that function as microclimates.

The ceramic pillars have empty chambers inside with iron pipes through which water circulates. This generates a flow that refreshes the atmosphere thus cooling the areas where queues are likely to form.
The effect of the water sprinkling represents a natural flow thus helping to reduce the summer heat that will occur in the building during Expo Zaragoza 2008. In addition to this, all the vertical air conditioning and electrical supply conduits are arranged in a tree-shape form. This has an important central block that enables vertical circulation and gives access to the levels where removable floors allow horizontal circulation. All exhibition spaces are dug into the ground. The forged elements are hung from the roof and braced only to the pillars that are wooden beams made from recycled shavings and resin. The exterior walls are made from glass and anchored to the structure. To achieve privacy and sound proofing, mobile partition systems will be installed. This will be necessary to hold some of the planned exhibitions and to use the site in the future as a cinema.

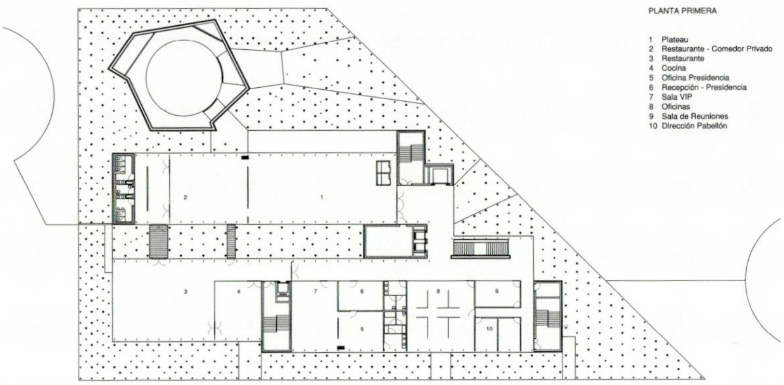

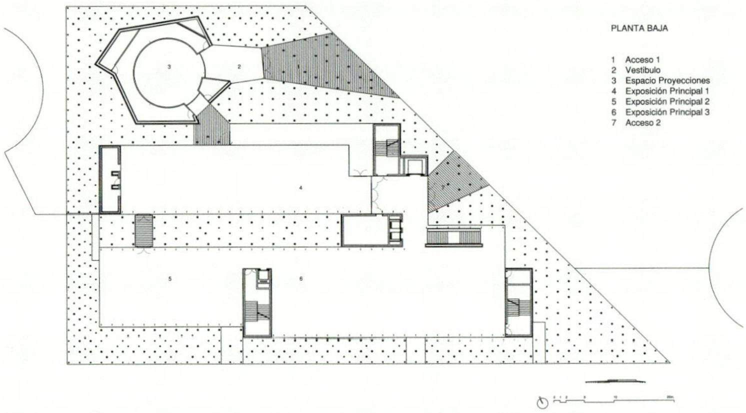

CONSTRUCTION DETAILS

01 Synthetic rubber roof waterproofing
02 Sandwich panel with waterproof veneer on upper face
03 Latticed framework for sloping roof
04 Concealed galvanised steel guttering
05 Folded galvanised steel facing
06 HEB 600 metal roof beams
07 Tubular purlin beam
08 Metal rib reinforcement: Variable height steel plate
10 Metal sheet
11 Composite slab floor
12 Mineral rock wool insulation
15 Waterproof membrane
16 Cork agglomerate thermal insulation
23 Internal facing of space with galvanised steel plate
24 Tongue and groove joining of MDF panels
25 Tiled steel pipe facing
26 Floor perimeter framework: Galvanised iron plates
27 Pine wood beam
28 Wood panel ceiling
29 Installation plenum
30 Metal plot for compact technical floor
31 Raised flooring with MDF panel base
32 Air conditioning heating inlet grille
33 Metal brace
34 Outer wall metalwork in galvanised steel
35 Thermal double glazing
57 Level floor
61 Reinforced concrete slab
62 Floor tiled in non-slip glazed stoneware
63 Swimming pool construction concrete layer
64 Waterproofing layer
63 Concrete slope
66 Infilling
67 Reinforced concrete foundation slab
68 Poured concrete layer
69 Gravel fill
70 PVC perimeter and basement wall drainage pipes
71 Wall waterproofing
79 Waterproof underwater lighting

CERAMIC PILLARS. CLADDING MANUFACTURING PROCESS

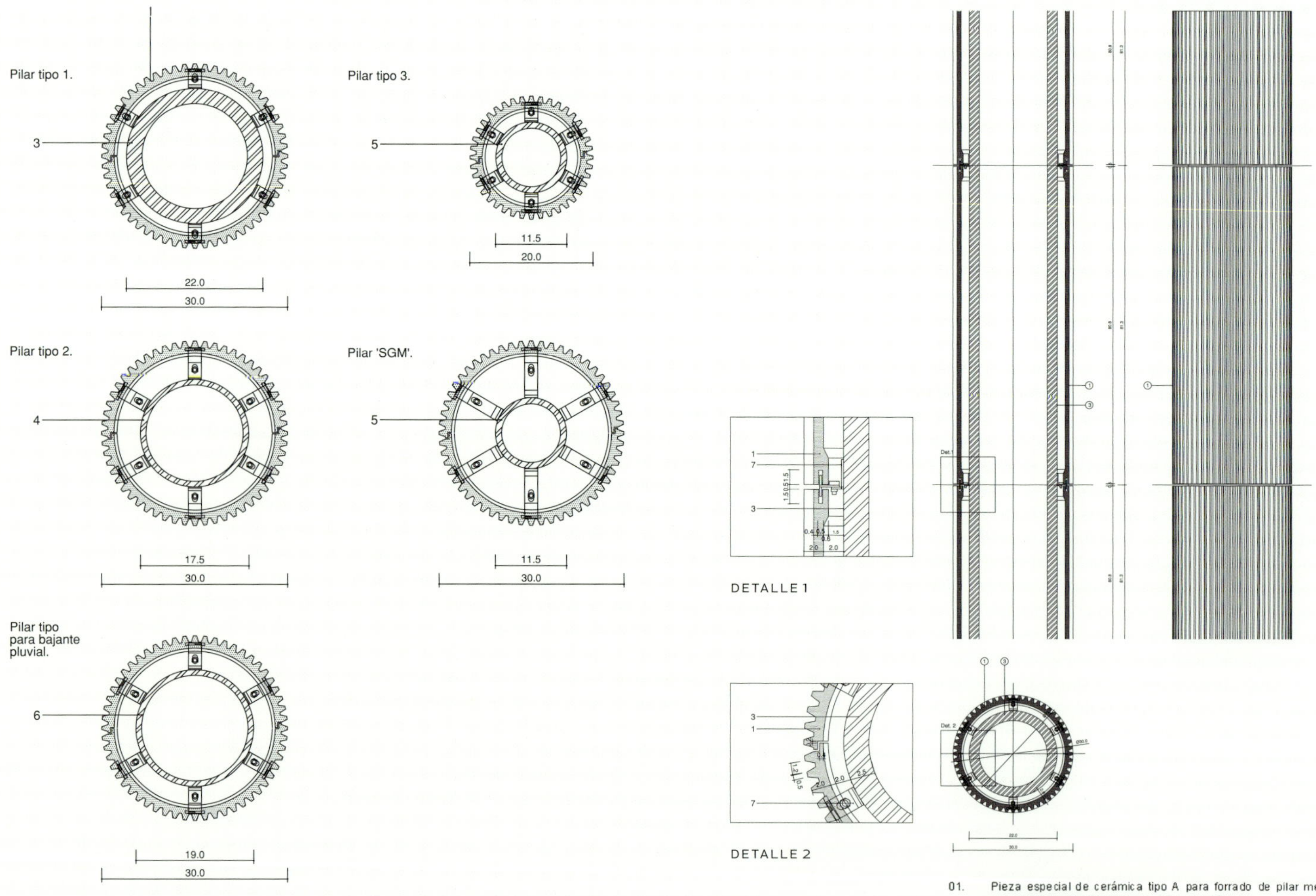

01. Pieza especial de cerámica tipo A para forrado de pilar metálico con acabado exterior crudo de 30cm de diámetro.
02. Pieza especial de cerámica tipo B para forrado de pilar metálico con acabado exterior crudo de 20cm de diámetro.
03. Pilar metálico 220.25mm.
04. Pilar metálico175.9mm.
05. Pilar metálico de 115mm.9mm.
06. Pilar metálico de 190.9mm.
07. Pieza para anclaje mecánico soldada a pilar y anclada a pieza cerámica.

CERAMIC PILLARS. BUILDING DETAILS

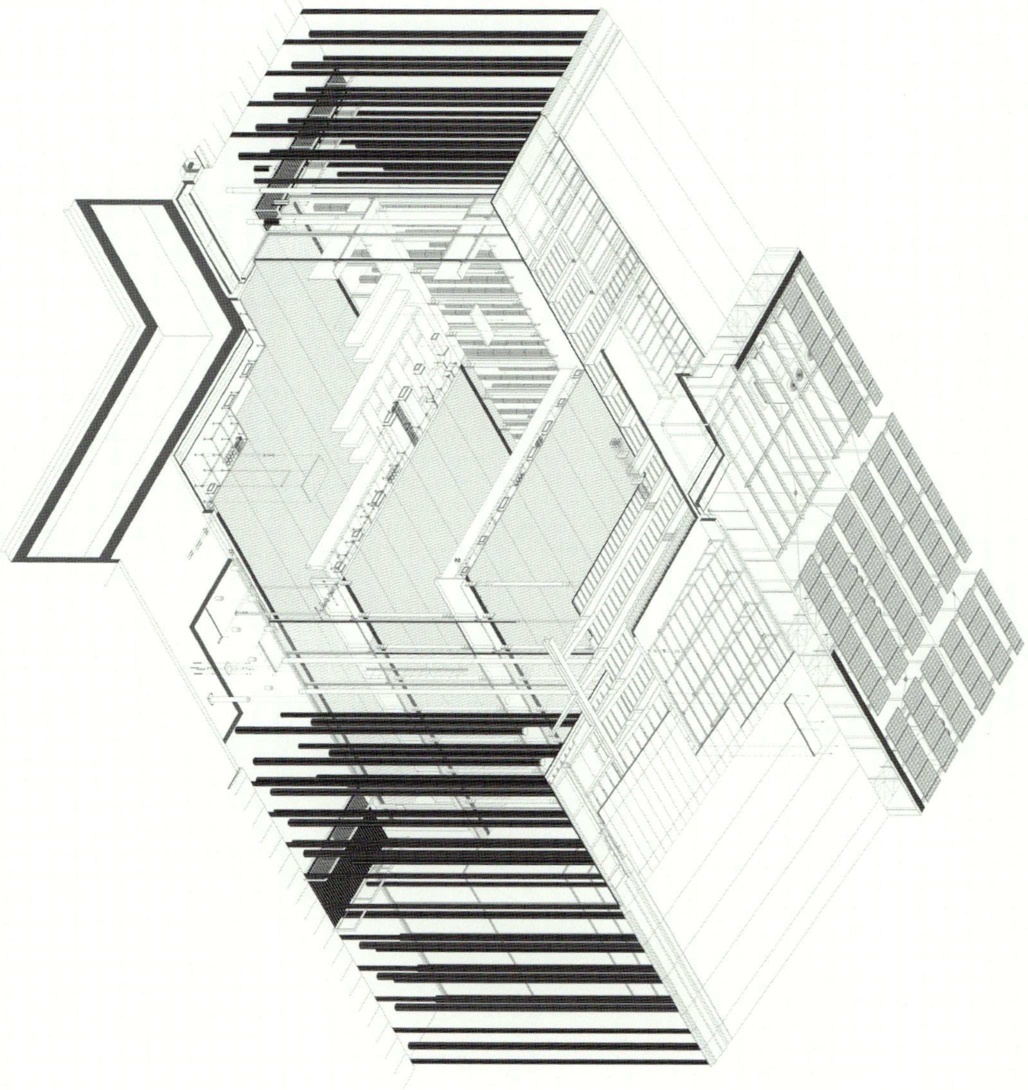

ARCHITECT: PATXI MANGADO / **LOCATION:** ZARAGOZA, SPAIN / **PROJECT:** DECEMBER 2005 – JULY 2006 / **CONSTRUCTION:** NOVEMBER 2006 – JANUARY 2008 / **COLLABORATORS:** CRISTINA CHU, JOSÉ M. GALSTALDO, RICHARD KRALOVIC, HUGO MONICA, ISABELLE PUTSEYS / **CLIENT:** SOCIEDAD ESTATAL PARA EXPOSICIONES INTERNACIONALES / **STRUCTURES:** NB35 / **BUILDER:** CONSTRUCTORA SAN JOSÉ S.A.

CERAMICS: ALL THE PILLARS IN THE BUILDING WILL HAVE A SPECIAL CLADDING AND WILL BE ASSEMBLED OVER A METAL CORE

Equip XCL
Xavier Claramunt

Ceramic Tiles in Motion Espacio ASCER Casa Decor Barcelona 2007

Barcelona, Spain

Ceramics actually move in this project, hence its name Ceramic Tiles in Motion. The project is formed by a quartet of motors, Opus 1, moving across spaces of between 1 and 3.5 metres. Music and imaginary mental spaces become a reality before our eyes as four 1,000 W motors with ceramic surfaces move at a rate of 1 m every 30 seconds. The two front motors oscillate across an area of 3.5 metres, while the two rear ones move across 1 m. Thus a cyclic, smooth choreography is commenced, always moving within the limits of the material. The motors play a game of seeking, waiting and overtaking to prove that ceramics do not have fixed points in space.
The tiled surfaces are made up of seven main sections measuring 92 x 10,000 cm. In turn, these are formed by sub-sections measuring 92 cm^3 in a 4+2+4 sequence. Nine sheets of tiles measuring 30.5 cm^3 are attached to each of these sub-sections. The sub-structure of all these main sections and sub-sections is formed by an electrically welded 5 cm^3 mesh with transversely welded handrails upon which the seven main sections are attached.
This mesh hangs from four hoists located 4.4 m above the floor, held by a steel frame with a total capacity of 6,000 kg. Each motor is automatically controlled and has an oscillation route ranging between 1 and 3.5 metres.

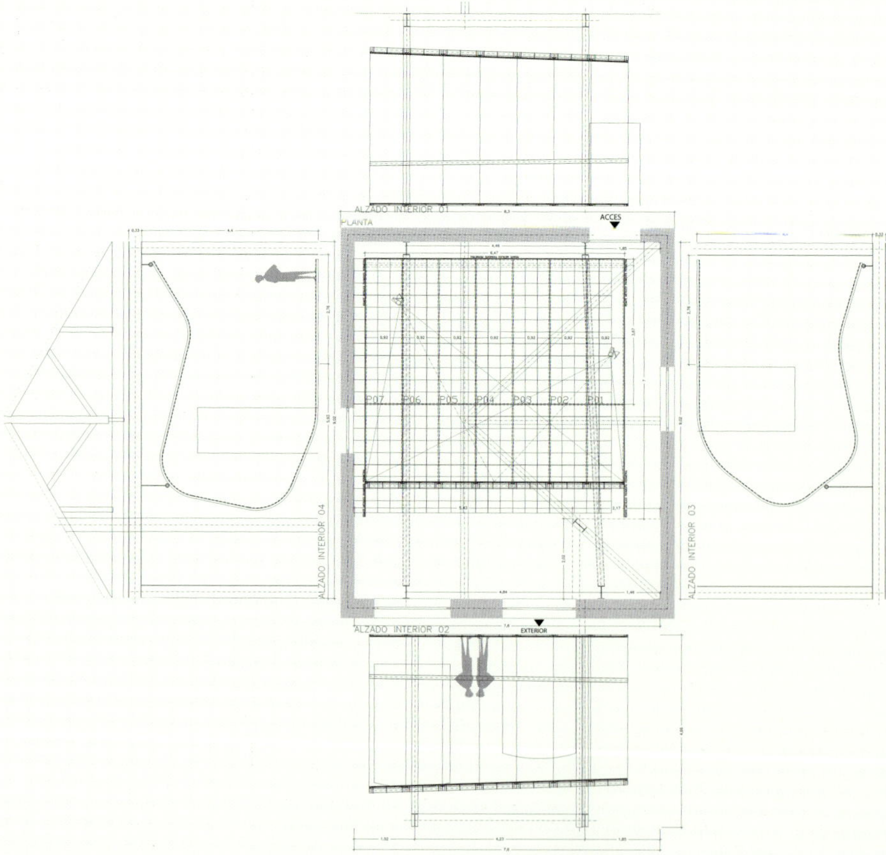

FLOOR AND ELEVATIONS

ARCHITECTS: Equip XCL. Xavier Claramunt / **LOCATION:** Antigua fábrica Fabra i Coats, Segre 30, Barcelona, Spain / **CHRONOLOGY:** 2007 / **CLIENT:** Ascer / **PHOTOGRAPHER:** Ainhoa Anaut

CERAMICS: The project is covered by ceramic tiles that rise and fold over themselves powered by four motors. The framework is formed by a 22.680 tiles and a total of 36.288 tiles are laid on the floor. Each tile is white and measures 50 x 50 mm

Carlos Ferrater,
Joan Guibernau

Casa Ubiñana

Sant Cugat del Vallès, Spain

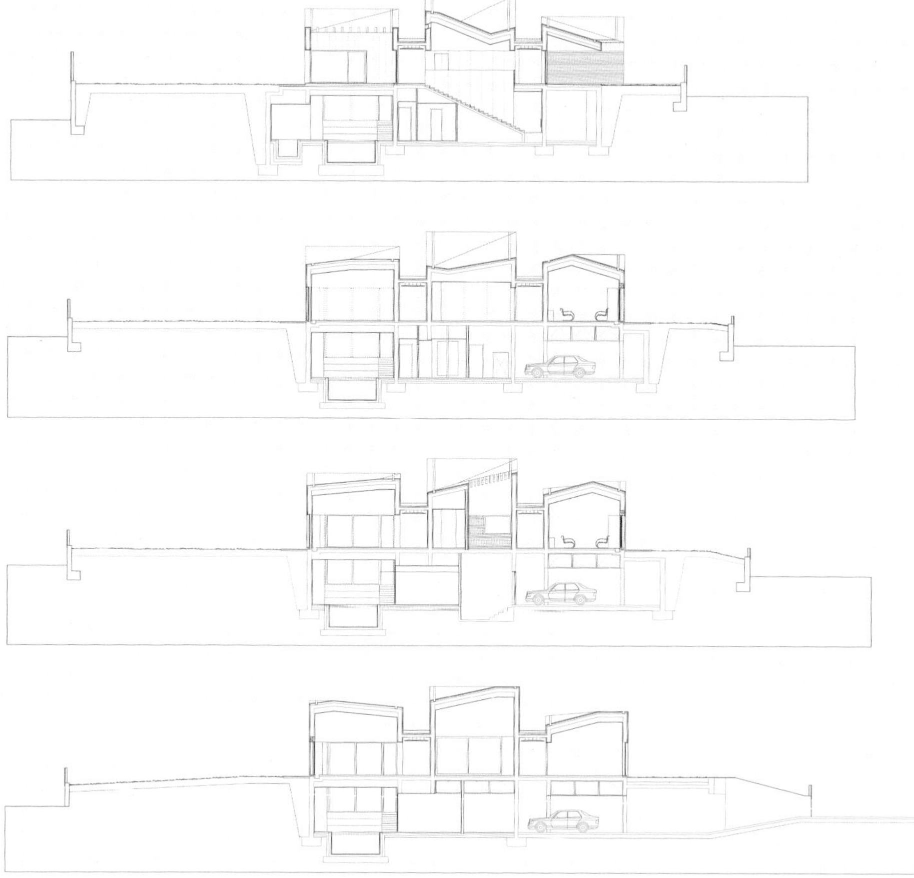

CROSS SECTIONS

The Casa Ubiñana house comprises three six metre high bands that contain the different rooms. The roofs are sloped in accordance with regulations and adapt to the unique interior. Their height adjusts according to needs of the interior space, creating a skyline that can be seen from the highest areas of the immediate surroundings.
Outside, flat green lines cover the various structures, connecting them to the garden. Inside, the different slopes of the roof planes form a series of linked spaces. Large white ceramic tiles were the only material to be used on the roof and façades. These were placed on battens to form a ventilated façade and are separated from the reinforced concrete supports that form the curtain walls and the roof slabs.
Small perforations were made in the horizontal planes between the bands to provide the inside areas with natural light. A heated swimming pool was placed in the basement, projecting onto the lower garden areas.
The car park entrance is in the lower area of the access road.

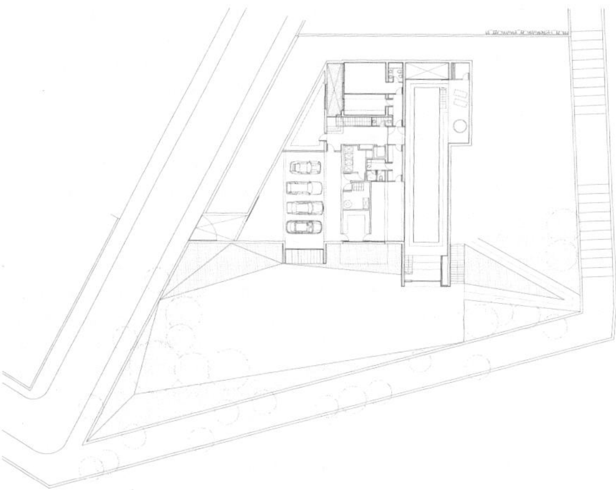

BASEMENT

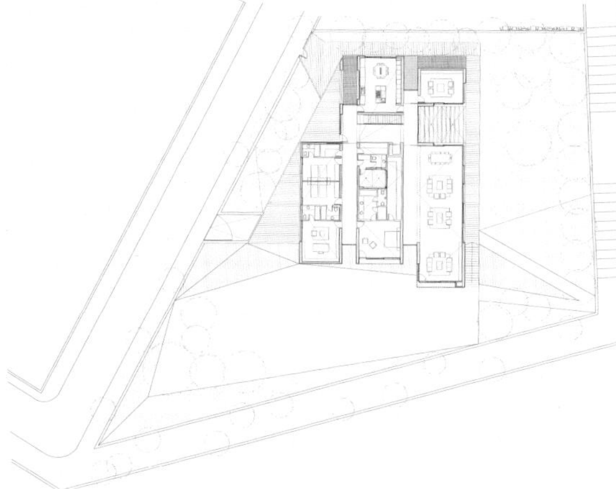

GROUND FLOOR

ARCHITECTS: Carlos Ferrater, Joan Guibernau / **LOCATION:** Carrer Adjutori Parcela 7 i 8, Sant Cugat del Vallés / **CHRONOLOGY:** 2003 – 2007 / **COLLABORATORS:** Sofia Machado / **STRUCTURES:** Juan Carlos Capilla / **BUILDER:** Pérez Villora / **BUILT AREA:** 1.000 m^2

CERAMICS: The façades and roofs are all covered with rectified porcelain tiles 450 x 900 x 10.4 mm (ivory, natural texture) assembled on metal battens and fastened to reinforced concrete walls and slabs

154

EMBT Arquitectes
Enric Miralles, Benedetta Tagliabue

Public Library

Palafolls, Spain

The project for the Palafolls library lasted a long time; the works actually finished ten years after the project was first envisaged. Owing to the many changes and the numerous people that contributed to the development of the original ideal, the results are exceptional.
As the project arrived on the EMBT studio desks in a moment when the company was undergoing strong growth and internationalisation, the library offered a chance to work on a different type of project and to develop a more intimate space.
The library design started with a single large room covered by domes – which had traditionally always formed a part of Catalan architecture. These are the same domes that Le Corbusier rediscovered for domestic architecture, as used in the villas at Sarabhai and Ahmedabad. The plans for the domes later became increasingly difficult, complicating the building of the library.
Once the Provincial Council had decided upon the winning entry, the decision was made that the design it should not be so self-absorbed; the idea was to create a more open relationship with the outside world.
The original structure therefore started to develop some most unusual shapes, with cut-off domes that still covered a single space.

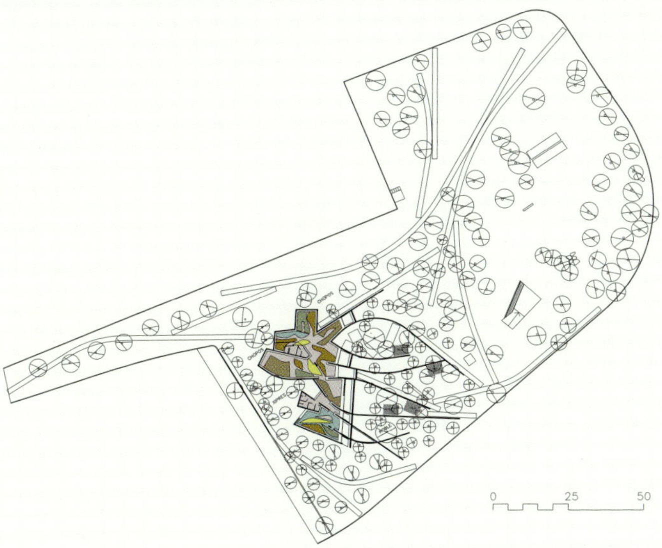

LOCATION

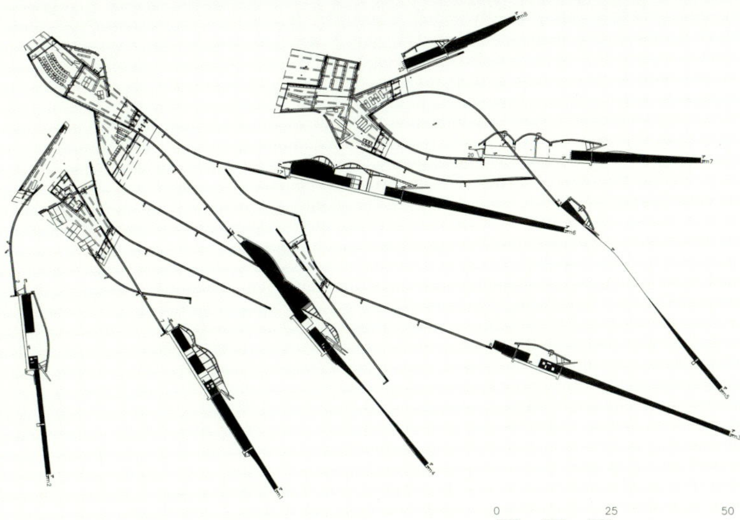

FLOORS AND ELEVATIONS

The walls transformed into lines that would guide people in from the outside and vice versa, forming patios where one could take a book outside and find a good spot underneath a tree to read it.
It is here, on the walls, where the ceramic material is found, used in a continuous manner as a quality covering, offering texture and guaranteeing good exterior protection against the elements.
The only real solid wall in the library is the one separating the inside from the outside.
This line is the library "door" and is a real mechanism for transforming the senses.

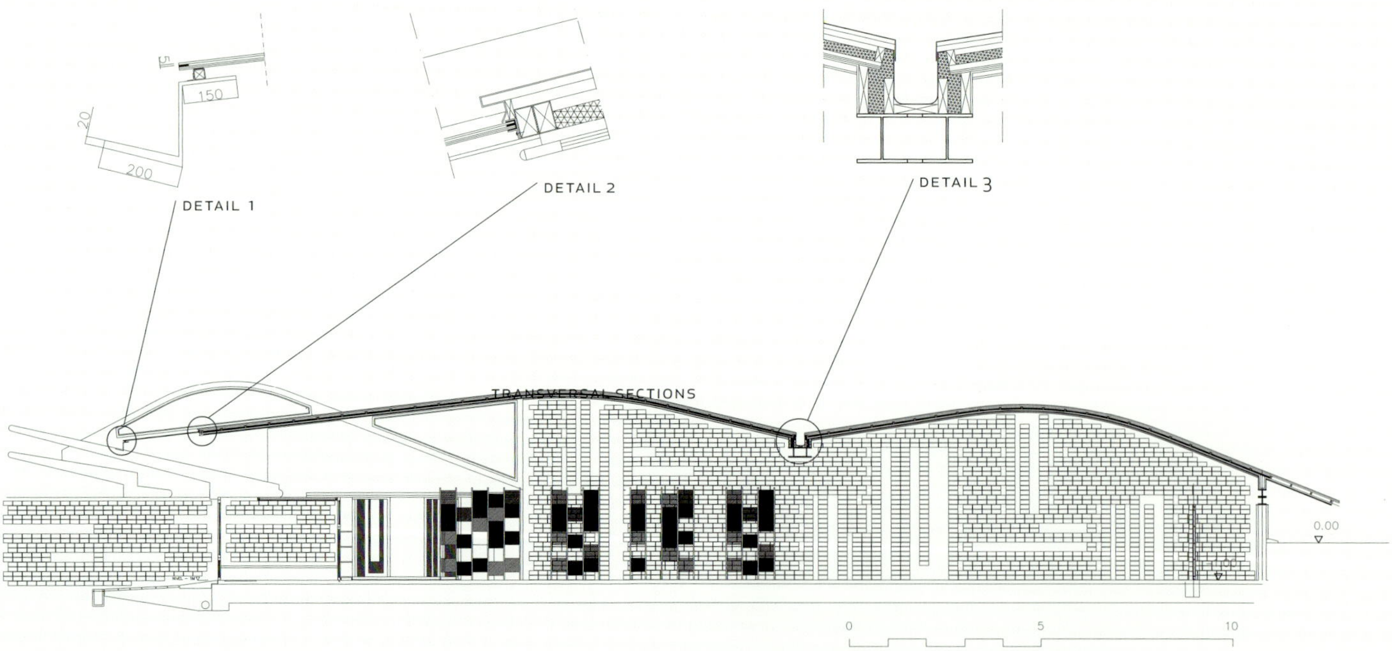

SECCTION AND DETAILS

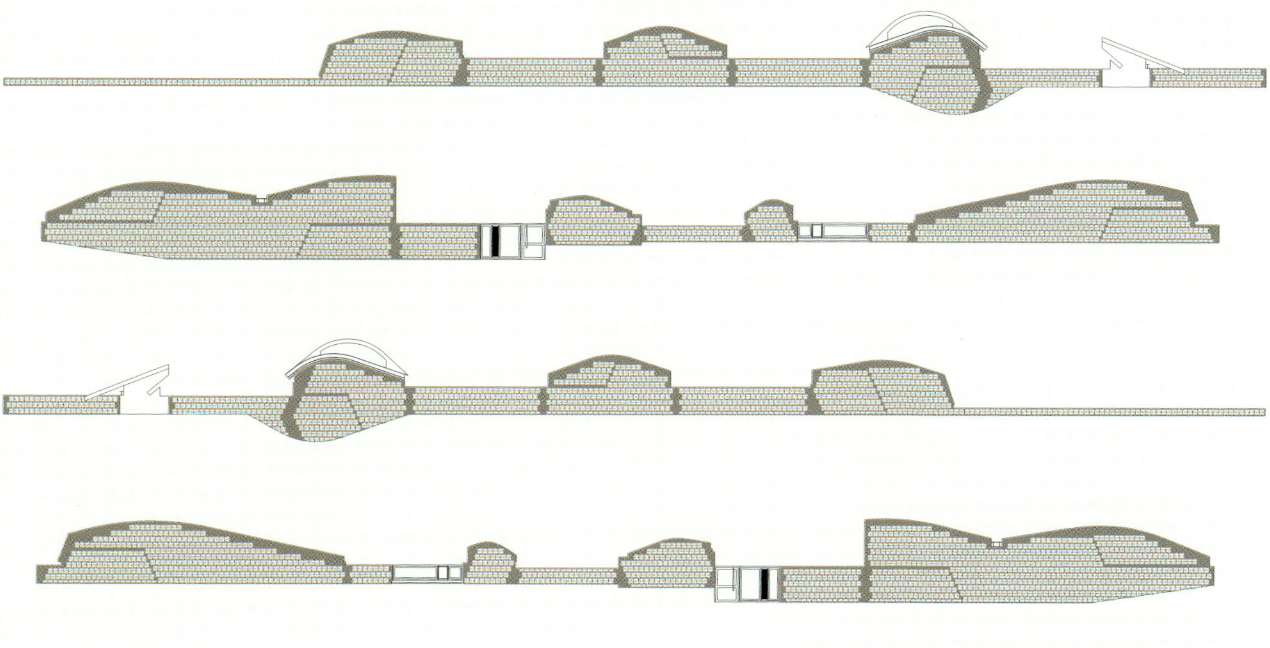

ELEVATIONS

ARCHITECTS: EMBT Arquitectes. Enric Miralles, Benedetta Tagliabue / **LOCATION:** Palafolls, Spain / **COMPETITION:** First Prize 1997 / **COMPETITION COLLABORATORS:** Elena Rocchi. Michael Eichorn, Niels Martin Larsen, Nicolai Lund Overgaard, Isabel Sambeth, Ana Maria Romero, Germán Rojas, Carlos Alberto Ruiz, Alfonso López, Marc de Rooij / **PROJECT:** 1998 – 2007 / **CONSTRUCTION:** 2000 – 2007 / **PROJECT MANAGER:** Josep Ustrell / **BASIC PROJECT:** Makoto Fukuda, Hirotaka Koizumi / **COLLABORATORS:** Guillame Faraut, Angel Gaspar Casado / **BASIC PROJECT:** Makoto Fukuda, Josep Ustrell / **COLLABORATORS:** Catalina Montaña, Lorena Gálviz, Marco Orozco, Daniele Romanelli, Ornella Lazzari, Marco Darío Chirdel, Tomoko Sakamoto, Sania Belli, Koichi Tono, Markus Lechelt, Jan Maurits Locke, Juan Carlos Mejía del Valle, Adrien Versuere, Richard Breit, FlorenciaVetcher, Jad Salhab, Nils Becker, Ezequiel Cattaneo, Manuela Schubert, Marco Orozco, Josep Miàs, Daniele Romanelli, Catalina Montaña, Ornella Lazzari. Natalia Leone / **CLIENT:** Palafolls Town Council / **ENGINEERING:** STATIC, Gerardo González, Barcelona, Nilo Lletjós (IOC) / **INSTALLATIONS:** PGI, Instalaciones Arquitectónicas / **BUILDER:** Brigada Municipal, Gives, Metálicas Valls / **BUILT AREA:** 714 m² / **PHOTOGRAPHER:** Duccio Malagamba, Cerámica: Brecor S.A.

CERAMICS: Baked solid ceramic pieces 29 x 14 x 4 cm used as interior and exterior wall coverings, slightly protruding (1.5 cm) from the vertical walls

Published by:
ASCER
Spanish Ceramic Tile
Manufacturers' Association

Editated by:
Giacomo Delbene

Essays:
Gonzalo Silva Moreno
Javier Mira Peidró
Mila Payá
Pich Aguilera
Teresa Batlle
Víctor Echarri
Vicente Sarrablo
Eduardo de Miguel

Authors:
Their authors

Translation:
Cillero & de Motta Traducción

Graphic Design:
Actar Pro

Production:
Actar Pro

Printing:
Ingoprint SA

Photographs:
Ainhoa Anaut
Estudio Ayape
Archea Associati
Angel Baltanás
Melissa Bauld
Marco Bruni
Lluis Casals
Toni Cumella
Uxío da Vila
Enzo Di Cristina
Estudio Duran
Alex Gaultier
Joan Guibernau
Huntley Hedworth
ILFA
Duccio Malagamba
Estudio Mangado
Ruth Meca
Jose María Molinos
David Morris
Pablo Orcajo
Eugeni Pons
Lluis Ros
Jaume Sala
Angel Sánchez
Eduardo Sánchez
Débora Tenenbaum
Estudio Tobías
Salvi Vivancos
Giovanni Zanzi

ISBN 978-84-612-1509-6
DL B-5920/2008

Printed and bound in European Union

For more information

ASCER
Spanish Ceramic Tile Manufacturers' Association
c/ Gínjols, 3
12003 Castellón (Spain)
tel: +34 964 727 200
fax: +34 964 727 212
global@ascer.es
www.spaintiles.info

ITC
Institute for Ceramic Technology
Campus Universitario Riu Sec
Avda. Vicent Sos Baynat, s/n
12006 Castellón (Spain)
tel: +34 964 342 424
fax: +34 964 342 425
itc@itc.uji.es
www.itc.uji.es

ALICER
Architecture and Design Area
Institute for Ceramic Technology
Avda. del Mar, 42
12003 Castellón (Spain)
tel: +34 964 342 424
fax: +34 964 342 425
universal@alicer.es
www.alicer.es

IPC
Institute for the Promotion of Ceramics
Avda. Vall d'Uixó, 25
12004 Castellón (Spain)
tel: +34 964 359 902
fax: +34 964 359 940
ipcdip@dipcas.es
www.dipcas.es

QUALICER
The forum on ceramic tile
Cámara de Comercio de Castellón
Avda. Hermanos Bou, 79
12003 Castellón (Spain)
tel: +34 964 356 500
fax: +34 964 356 510
qualicer@camaracs.es
www.qualicer.org

CEVISAMA
International trade show for ceramic tiles, surface coverings, bathroom and kitchen equipment, raw materials, glazes, frits and machinery (machinery only even years)
Feria Valencia - Cevisama
Avda. De las ferias, s/n
46035 Valencia (Spain)
tel: +34 902 747 254/5/6 -
+34 963 861 185 - +34 963 861 377
fax: +34 902 747 257
cevisama@feriavalencia.com
www.feriavalencia.com/cevisama

Tile of Spain Center - Germany
Spanish Commercial Office in Düsseldorf
Jägerhofstrasse, 32
40479 Düsseldorf (Germany)
tel: +49 211 493 660
fax: +49 211 499 711
dusseldorf@mcx.es

Tile of Spain Center – United States
Spanish Commercial Office in Miami
2655 Le Jeune Road, Suite 1114, Coral Gables
Florida 33134 (USA)
tel: +1 305 4464 387
fax: +1 305 4462 602
miami@mcx.es

Tile of Spain Center - France
Spanish Commercial Office in Paris
11, Avenue d'Iéna
75016 Paris (France)
tel: +33 1 53 579 550
fax: +33 1 47 209 722
paris@mcx.es

Tile of Spain Center - Italy
Spanish Commercial Office in Milano
Via del Vecchio Politecnico, 3-16°
20121 Milán (Italy)
tel: +39 02 781 400
fax: +39 02 781 414
milan@mcx.es

Tile of Spain Center – United Kingdom
Spanish Commercial Office in London
66 Chiltern Street, 2nd Floor
London W1U 4LS (UK)
tel: +44 20 74 672 330
fax: +44 20 74 875 586 / 72 246 409
londres@mcx.es

Tile of Spain Center - Russia
Spanish Commercial Office in Moscow
Ulitsa Vozdvízhenka, 4/7 (entrance from Mokhovaya)
125009 Moscow (Russian Federation)
tel: + 7 495 7839 281/2/3/4/5
fax: +7 495 7839 291
moscu@mcx.es

ICEX
Spanish Institute for Foreign Trade
Paseo de la Castellana, 14-16
28046 Madrid (Spain)
tel.: +34 902 349 000
www.icex.es

With the support of

ICEX
SPANISH INSTITUTE FOR FOREIGN TRADE